Research Project Success
The Essential Guide for Science and Engineering Students

Research Project Success
The Essential Guide for Science and Engineering Students

Cliodhna McCormac
NIBEC, School of Engineering, University of Ulster, UK
E-mail: c.mccormac1@ulster.ac.uk

James Davis
NIBEC, School of Engineering, University of Ulster, UK
E-mail: james.davis@ulster.ac.uk

Pagona Papakonstantinou
NIBEC, School of Engineering, University of Ulster, UK
E-mail: p.papakonstantinou@ulster.ac.uk

Neil I Ward
Department of Chemistry, University of Surrey, UK
E-mail: n.ward@surrey.ac.uk

RSCPublishing

ISBN: 978-1-84973-382-3

A catalogue record for this book is available from the British Library

Published by The Royal Society of Chemistry,
Thomas Graham House, Science Park, Milton Road,
Cambridge CB4 0WF, UK

Registered Charity Number 207890

For further information see our web site at www.rsc.org

Printed in the United Kingdom by CPI Group (UK) Ltd, Croydon, CR0 4YY, UK

Preface

A degree course typically lasts three years and can occasionally last four. Throughout those years, you slog away attending lectures, tutorials and lab sessions – dutifully completing one assignment or report after another. More than likely, these will be the same assignments and reports that were set in previous years. Then, finally, you get to the exciting part – the project. And that's when, all the knowledge you've accumulated and all the skills that have been so carefully honed should come together and you can venture into the world of research. This is where you get to put your stamp on the knowledge base that drives humanity forward. Someday, someone will be writing an essay or completing an assignment on the work that you pioneered. Well...that's the theory. Unfortunately, that cherished dream of pushing back the frontiers of science can easily become tarnished – not for want of trying – but through failing to know just what to do or how to go about it. This can sound a little silly, given the years of preparation that have gone into getting you this far, but it is an immutable truth that a project requires the development of a whole new set of management skills that have hitherto been hidden. Our purpose in writing this book is to uncover those skills and develop them to the point where you have the confidence and know-how to tackle the project head on.

Research Project Success: The Essential Guide for Science and Engineering Students
Cliodhna McCormac, James Davis, Pagona Papakonstantinou and Neil I Ward
© Cliodhna McCormac, James Davis, Pagona Papakonstantinou and Neil I Ward 2012
Published by the Royal Society of Chemistry, www.rsc.org

The majority of lab sessions will involve following a recipe of some form or another and effectively take you by the hand through the procedures. In contrast, the project is where you are expected to lead – well at least a bit. You are given initial instructions, but then the onus falls on you to supply the enthusiasm, the ingenuity, and the professionalism to ensure it goes the distance. This is not to say that you are left all alone to decide what to do. It is imperative that you make a contribution and play your part. At the end of the project you must produce a professional report that will impress not only an examiner, but also, potential employers. When you create something new – there is always a dilemma as to how people will receive it. Does it have all the bits necessary? What are those bits? Are those bits in the right order? Where do you start? What should you do first? How do you go about getting data? What do you do with the data once you get it? How do you write up the data? These are key questions that can flummox any student.

This book does not aim to be a conventional textbook with a rigorous outline of all the things that must be done. Rather, it is meant to be a companion who tries to give you a helping hand, to guide you safely through such seemingly treacherous terrain, to provide you with an overview of the actions you need to take, from selecting a project through to the final write up and poster or oral defence of the dissertation. The book does not give an in depth description of how to do the nitty gritty of the project – it cannot – projects are too varied and any such book would be unwieldy and you certainly wouldn't want to read it. What it does do – is gently nudge, prod and occasionally shove you in the right direction. It serves to reinforce the key management skills you need to develop and to give you the confidence needed to pursue the various aspects of the project effectively. Many projects are let down by the fact that the student is left unsure as to what is expected of them and what should go into the dissertation. The following chapters take each part in turn, considering the problems and suggesting how best they can be overcome. It highlights where students commonly go wrong and provides invaluable tips to help you gather as many marks as possible. While pushing back the frontiers of science is very laudable – it is best to think, at least initially, in terms of submitting a really good dissertation that will gather you a slew of marks and contribute meaningfully to your degree classification. It is our hope that the suggestions within this book will remove some of the stress that can arise and allow you to actually enjoy the project.

CMcC, JD, PP, NIW

Contents

Research Project Success: The Essential Guide for Science and Engineering Students
Cliodhna McCormac, James Davis, Pagona Papakonstantinou and Neil I Ward
© Cliodhna McCormac, James Davis, Pagona Papakonstantinou and Neil I Ward 2012
Published by the Royal Society of Chemistry, www.rsc.org

CHAPTER 1

First Steps in an Epic Journey

1.0 THE RESEARCH PROJECT

Picture the scene: Christopher Columbus poised on the deck of the Santa Maria, about to set sail on a voyage to forge a daring new trade route to the East Indies. His fingers tap absentmindedly on

Research Project Success: The Essential Guide for Science and Engineering Students
Cliodhna McCormac, James Davis, Pagona Papakonstantinou and Neil I Ward
© Cliodhna McCormac, James Davis, Pagona Papakonstantinou and Neil I Ward 2012
Published by the Royal Society of Chemistry, www.rsc.org

the bow of the ship as he contemplates the journey ahead. He feels both excitement and trepidation, in equal measure – after all, this will be the first time he has led such an expedition. Unlike previous sailings – where he was simply one crew member among many following established trading routes – he now has the resources at his disposal to pursue something new and potentially magical – effectively, to pursue his own dream – but, its success lies on his shoulders. He possesses the basic sea-manship training along with a working knowledge of the sea and the requisite navigatory skill. He also has a rough idea which way to go – at least he thought he did (maybe navigation wasn't his strongest subject). As he looks around the ship, he sees a crew of able-bodied men who will help him make this voyage happen. Granted, this is a somewhat romanticised picture, but it embodies the very essence of a research project and, irrespective of your specific discipline, there is little doubt that you will experience similar feelings to Columbus – minus the fear of shipwreck or death by sea monster, of course.

The research project is undoubtedly the most exciting part of the degree programme as it is where you get to put into practice the skills and knowledge that you have acquired through countless lectures, labs and tutorials. As with Columbus, you are about to embark on a journey with the assistance of a dedicated research team, a journey that aims to pave new ground that could help to confirm a hypothesis and may even lead to some startlingly unexpected discovery. It makes no difference as to the research area you are about to become a key part of, the project is very much an expedition into the unknown and one that needs careful planning as it can also be the most fraught part of the degree programme. The voyage undertaken by Columbus was far from plain sailing and there would have been many times where things weren't going quite to plan. This book will help you plan your expedition so that the hard times can be weathered and also provide you with a toolbox that will enable you to tackle the ups and downs that can arise on the way.

1.1   WHICH DIRECTION TO TAKE?

First, you have to choose a project – but which one? The proce-dures for allocating projects will vary from one university to another but, in general, most tend to involve the circulation of a list of project titles and then it is up to you to choose your favoured three or four. Depending on the organisation of the particular degree programme, this could either be at the end of the second year or upon returning to your final year. The list itself will reflect the various research interests within the department and will be associated with one or more supervisors. There is an increasing trend toward interdisciplinary projects (Box 1.1) and it is not uncommon to find that the supervisory team is made up of two academics – sometimes from different departments or even from different institutions or commercial organisations.

BOX 1.1

Interdisciplinary is an increasingly used key word and it is vital that you can demonstrate to potential employers that you:

- have a flexible skill set;
- can communicate across a number of levels;
- can easily integrate within a team.

The involvement of another branch of science, even if it is one that you are not familiar with or have previously shied away from, shouldn't be viewed as a bad thing and should not be needlessly discounted as it can broaden both your outlook and, importantly, your employability. As you go through the degree programme you will be aware of where your strengths lie and equally those subjects in which you are weak. It would be wise to avoid projects that are predominantly composed of the latter. When selecting possible projects it is best to play to your strengths. It is important to remember that you only have a very limited amount of time to conduct the practical work and you should be aiming to give yourself a greater chance of picking up marks rather than using the time to master a subject that has hitherto been a struggle.

While the mechanics of the project allocation will vary, most institutions employ a practice where you select a number of titles and rank them in order of preference as there will be an almost inevitable tendency for some titles to prove more popular than others. Obviously, not everyone can do the same project so there must be some careful management to allocate students evenly across the department whilst still attempting to cater for at least one of your preferred choices. Therefore the project list you submit becomes crucial and is not something that should be done on a whim (Box 1.2).

BOX 1.2

There are a number of considerations to be aware of at this crucial stage in planning your expedition and these relate to:

- the lecturer(s) involved;
- large *vs.* small research groups;
- resources available;
- easy or trendy projects.

1.2   LECTURER/SUPERVISOR PERCEPTIONS AND MISCONCEPTIONS

Columbus didn't sail alone. He had a crew whom he picked to help him on the voyage – people in whom he could confide, ask for advice and who were more familiar with the sea and its ways than he was. He was surrounded by people who possessed the key skills needed for the day to day rigors of ensuring everything was ship-shape for the journey. This mirrors any science lab – where you have academics, technicians, administrators and a host of post-graduate and postdoctoral staff all working to ensure that the research activities run as smoothly as possible. Given the duration and significance of Columbus' expedition, it would have been unwise to randomly pick the crew. Likewise, it would have been incredibly silly to simply invite all his friends on the basis that they were nice and could tell a good joke. You must now take a critical look at the projects on offer, but also the people that advocate and support them.

Throughout your studies you will have come across many different lecturers – each with their own quirks. It is very tempting to think that because a particular lecturer is nice, friendly, etc. that they would be good to work for and therefore their project title is automatically the one to go for. This could easily lead to disaster as you may end up doing work that you don't actually like – or even worse – detest. There is no doubt that the working relationship between you and the supervisor is an important consideration, but it is only one of many (Box 1.3). The gruff lecturer whom you might not have liked that much in lectures, may be doing much more interesting and rewarding research and it would be extremely unwise to simply discount their value based on previous encounters.

BOX 1.3

Prospective supervisors:

Good Indications
- enthusiastic;
- clear plan for project;
- publication record.

Bad Indications
- lack of real interest;
- project has run unchanged for years;
- vagueness.

It is important to appreciate the difference between an academic's role as a lecturer and that of a project supervisor. Admittedly some may see or express little difference between the two but most academics come alive when conducting research and, increasingly, it is the outputs from the latter that are used by the universities to gauge their performance and indeed success. You may find that the attitude of a lecturer can change quite dramatically when considering research as opposed to supervising labs that have run unchanged for decades. You should find that they have a passion for their particular subject and especially for the projects they are offering. In many cases, they will be desperate to have someone share in their enthusiasm and help them take the work forward. This should be an early indicator of a potentially **good supervisor**. Someone who conveys little or a vague interest in the subject is unlikely to motivate you or provide the time when things may not be going so well.

1.3   RESEARCH GROUPS: LARGE AND SMALL

Be part of a team not
a clone

The general attitude of the proposed supervisor is clearly important, but it is also vital that you gauge the level of support that they
can give you in terms of time and resources – contact time, access
to equipment and mentoring in terms of developing the key skills
needed for the project. It can be tempting to look at large research
groups and think that they are successful, as they are doing cutting
edge work and that you will get to use the latest equipment. This is
perfectly plausible, but there is an inherent danger that after being
integrated within a large team you may become disillusioned and
feel as if you have become little more than a small cog in a much
larger machine. That may well be the truth, however, it is not
true to assume that the contribution you will make will be mean-
ingless. Most projects, especially those for which the supervisor has
external funding, will have goals that may take many years to
realise and every bit of work will count. It is easy to get caught up
in the glamour of the title and lofty end goals, but be prepared for a
diet of routine hard work punctuated with moments of glee when
something has worked out rather than a daily whirlwind of glamour and Eureka moments as peddled by Hollywood.

A core advantage of opting to work within a large group,
however, is that there will often be an extensive support network
through the postgraduates or postdoctoral researchers within the

lab and it is likely (almost inevitable, in fact) that one of them would be given the express instruction to 'look after' you. While this is quite a traditional approach and often a management necessity for supervisors with very large groups, it is important to ensure that the researcher is not your sole source of support and that you still have regular contact with the supervisor.

It is also important that you should not discount those supervisors who have more modest groups or even those who appear to be 'lone rangers'. It would be extremely unwise to assume that they have little to offer and that by doing a project with them you would be at a disadvantage when compared with students pursuing their projects in large teams. An argument can be made where there is the potential for greater student–supervisor interaction within smaller groupings – it can be much more focused and, if driven by an enthusiastic supervisor, the projects could have more direct relevance or impact than one within the larger teams where there may be a tendency towards following a more prescriptive project/ approach with only an incremental (though no less valuable) contribution.

1.4   WHAT RESOURCES ARE AVAILABLE?

What is your role? Well what we
had in mind was...

The provision of adequate resources is another critical question that needs to be considered before selecting your project and embarking on your expedition. It would be a foolish explorer who sets sail without the appropriate equipment, equipment that is in working order or sufficient rations to see them through the journey. This is very much true of your situation. It is vital that you discover what the situation is with respect to access to equipment and whether or not the reagents, materials or samples are available. The latter is particularly pertinent where the project involves the examination of biological, environmental or industrial materials from sources outwith the university. It is imperative to check what arrangements have been made – are they already in place or do they still need to be collected? Is there a backup plan should the samples fail to arrive? Is ethics approval necessary and, if so, has it been granted? These issues affect research groups irrespective of size.

Access to equipment and facilities (such as fume hood and bench space, etc.) can be a source of contention within many labs and it is important that you establish where you will be working and what equipment you will be expected to use and your 'right of access' to it. The situation gets more complicated as the expense of the equipment increases and where they are utilised as a central resource for an entire department. This is not so much an issue if, in your project, such equipment is likely to be used sparingly (i.e. for the one-off characterisation of a sample you may have prepared), but it does become problematic if it is core to your project and where other people have equal or greater rights of access. The sharing of resources is an inevitable consequence of the high cost of equipment and is not to be taken as something that automatically precludes the selection of a particular project. In many cases there will a booking regime or rota. Providing the equipment is well maintained then there will seldom be an issue and all that will be required is a little compromise. One situation to be wary of, however, is the reliance of a project on antiquated equipment that has a history of downtime – whether age related or due to the absence of appropriate technicians or consumables to allow it to function. If it fails for any reason and is out of action – is there a viable alternative?

In some cases, such as a literature survey or computational projects, consumables and materials may be less of an issue, but the provision of resources remains a key determinant that must always be considered before selecting the project. In such cases computer access and the availability of the appropriate software or licenses may well be the crucial issue. Is there off-campus access or is the software restricted to a dedicated lab or computer system within the university? If the latter, who else uses it?

1.5   PROJECT TITLE LIST – THE SUPERVISOR'S SALES PITCH

Don't be seduced by the project title

In some cases the list may simply contain the title of the proposed project and the supervisor associated with it. In others, there may be a small paragraph explaining the aims of the project and perhaps some brief background. It is likely that even before casting your eye over the titles you will have an idea of who you would like to work with – the previous sections should have made you think twice and you should look at each project on its merits. There are still some things to be wary of with respect to the projects themselves. You should NOT simply select a project on the basis that you think it is going to be easy and therefore your life in the last year of the programme will be plain sailing. You have to remember that the project is one of the key discussion points in any job interview and it can often be a selling point. It is also likely, that the 'easy' looking project may well be anything but easy and involve long tedious hours trying to get data that has real-world meaning.

This brings to the fore another issue – projects that have changed little year on year. Again, there is a temptation to draw comfort from the fact that previous students have done it and passed and therefore you will too. The question you have to ask is how does the project you are about to embark on differ from theirs? Few people would have been impressed if Columbus had suggested an epic voyage round Gibraltar.

At the other extreme is the project title that implies cutting edge and sounds incredibly complicated. Are you up to the task? The answer to that question is yes. That is providing you have checked out the issues noted in the previous sections of this chapter. In all honesty, the trendy title is usually put forward as an attempt to capture your imagination and, if it has, then it may well have the same effect on a prospective employer. However, you may find that behind the façade of the fancy jargon, the actual work is routine.

Consequently, the project title should only be used as a gauge as to the general area in which you wish to participate. The crucial step involves you doing the forensic investigation into what it actually entails. You have to question everything about the project and have no fear about approaching the supervisor or other members within their team to find out the actual facts. The project may paint a picture of pursuing a goal that is going to lead to a major scientific breakthrough but, in reality, it is unlikely that you will be able to accomplish that within the timescale allocated to your project and the work may be a small but nevertheless important step forward for the research team. Therefore, no matter how the title is dressed up – whether to make it sound easy or enthralling or your doorway to the next Nobel Prize – you actually need to do some serious legwork to find out exactly what is involved before you make the final decisions and commit your selection to that all important form.

1.6 INTERVIEWING YOUR TEAM

Prepare a list of questions before approaching
a prospective supervisor

We have established that the worst thing you can do is sit down with the list and simply pick those projects that you think sound nice or easy or where you merely like the style of the lecturer. So what do you do? The answer is simple. You have to be proactive and effectively interview the supervisors of the projects that you are interested in. It is unlikely that Columbus would have simply hired sailors on the basis of them proclaiming to have seafaring skills.

He would have demanded more details and the same is true for you. This may sound a bit intimidating, especially if you are not confident in approaching staff members. However, it is important to remember the significance of choosing the correct project and that must drive you to overcome any reticence.

The project will invariably count for a large chunk of your final mark. You will spend a fair amount of time completing it and, at the end of the day, you need to have a dissertation that you would want to show to prospective employers. It is therefore essential that you do everything within your power to make sure you have the best possible start. Getting that start means that you have to put in the effort to find out as much as you possibly can prior to making that all-important decision. If you just pick willy-nilly, there is a good chance you will end up with a project that you don't like, that wasn't what you thought it would be and where there may be little or no support to guide you through.

So what are the first steps? Once you have the list – go through it carefully and mark those project titles that you think sound the most interesting to you (avoid thinking about all the other stuff – just concentrate on what catches your imagination). Once you have annotated the list – narrow it down to about four or five preferred choices. Now the next task is to find out a bit more about them. Before running off to chase the respective supervisors – do some quick background reading on the subject. This could be a simple Google search or a browse through Wikipedia. At this point, you are simply looking to get a brief overview of the projects. You should then do a quick search of the supervisors publications on the subject – these will usually be listed on the university web site though the list can be out-of-date depending on how efficient the supervisor/department is in updating the web data. A better approach would be use the ISI Web of Science database (discussed in detail in Chapter 3) as this will give the most recent publications. These will give you an indication of the frequency with which the supervisor publishes and whether they have an extensive background in that particular area or whether they are using the project to investigate a new avenue.

This is almost akin to an academic credit check on your prospective supervisor as it lets you see how active they are in terms of research and should also give you an indication as to the possibility of your work ending up in a publication. If they publish frequently – have there been any undergraduates cited in the author list? This is another consideration and an important question to ask. If possible, download the articles and skim through them to get a feel for the work/techniques that have been used as these may, in all likelihood, be similar to the work that you would be doing. Arguably, the most important section of the papers at this point will be the introduction sections as it is here that the author – your potential supervisor – will effectively set out the aim of the investigation and justify the reasons behind pursuing that particular method. The significance of doing this background reading is that it gives you a more rounded picture of the type of work that the project is likely to entail and why it is being done. It also has an important side benefit in that when you go to speak to the supervisor about the work – you will be prepared when they ask you "are you familiar with…." This is where the supervisor is checking to see how keen you are. At this stage, it is obviously good if you can demonstrate that you have done the background reading and the best way is to answer by saying "Yes…I read/saw that it in……."

If, after this background check, you are still interested in the project then the next step is to email the prospective supervisors to arrange a meeting. This is a much better approach than simply turning up unannounced at their office or hounding them before or after a lecture. In arranging a fixed time for a meeting, the supervisor is less liable to feel harangued and should have time to explain about the project. It may also allow them time to show you the labs and equipment that would be used. Moreover, it is the time when you get to see the supervisor in a more relaxed setting where they can discuss the project freely and hopefully enable you to ascertain, on a personal level, what it would be like to work for that particular person. It is at this point that you may see the nominally 'gruff' supervisor display hitherto concealed enthusiasm and an approachability that may be seldom seen in the traditional lecture or lab. The meeting is not, however, simply an opportunity to have a vague chat about the project – it must be carefully choreographed by you to extract the maximum amount of information in a short time frame. Thus, you must go in armed with a list of questions about the project as you are effectively interviewing them.

1.7   MEETING CHECKLIST

When all is said and done and the facts are analysed
...go for the projects that stir your imagination

☑ What is the overall aim of the project?
You should already have a rough indication as to the direction that the supervisor wants to take from your earlier detective work, but sometimes the significance in the real world may not be clear.

☑ What do you expect this particular piece of research to achieve?
In many cases the project will be part of a larger research effort. This allows you to probe what is expected of you and what your contribution will be. Is this a follow-on project or a completely new area of research? Is there a possibility that it will lead to publishable results?

☑ Are there any other collaborators involved?
Is this a sole effort on the part of the supervisor or are there other departments, government agencies or commercial bodies involved? If there are other collaborators, what is their relationship to your project? This could be an opportunity to find out about samples and equipment, etc.

☑ How many people are involved?
Are there other postgraduates, research fellows or technicians involved? Are there any other undergraduates doing similar projects? There may well be several students doing a similar project. This question therefore allows you to ask what the difference will be between the projects. If the project has been running over several years – what was the outcome of the previous studies? Have the results been presented or published?

☑ What techniques would you use?
This will give you an idea of the skill set that will be developed. Are you going to be stood in front of one machine running endless samples or will the work be more varied? This enables you to get a feel for what you would be doing on a day-to-day basis. It also allows you to probe the availability of the equipment and your access to it.

☑ What would the supervision arrangements be?
This is where you discover the commitment of the supervisor. Will you be taken under the wing of a postgraduate or a postdoctoral researcher or do you report to the lecturer? How often would you meet up with the supervisor? Do they operate an open-door policy? Who will guide you on the practical aspects?

☑ Can you show me where I would be working?
This is THE question. In most cases, the lecturer will probably show you where you would be working and it may well be that you end up having the meeting in the actual lab where they do their research. If not, however, then you MUST ask to see where it is you will be based. Why is this so important? You will spend a considerable amount of time there in the coming months and it is therefore vital you know what the environment is like. It is easy to paint a picture with words, but sometimes peoples perceptions differ and what is brilliant, vibrant and exciting to one person maybe induce a feeling of terror in another. It is a bit like buying a car or a house. You would seldom go on the wording of the advert or the buyer's recommendations. You have to see it for yourself and decide for yourself

whether it is right for you. If you are not sure about any aspects, now is the time to ask.

☑ Is there any information you can give me to take and look at? You should ideally let the lecturer know that you have done the legwork and already have some information on the background to the project before asking this question. It is likely that most lecturers will simply refer you to their publication list and if that is the case then at least it gives you a way of checking that you have the relevant papers. If the project is a new area for the lecturer, they themselves may not have published on it, but they can then direct you to appropriate papers. At the very least, providing you have already demonstrated that you have done some forensic legwork, it will highlight your enthusiasm for the project.

1.8   RANKING THE PROJECTS

Unfortunately, there is no magical formula for deciding which one to pick. The question checklist provides you with the facts on each one, allows you to compare them and gives you the reassurance that all the right factors are there. It could be that some appear deficient in some areas and excel in others – YOU have to weigh up the pros and cons and decide which are important to YOU. If there is a question over obtaining samples, then is that such a big deal when you may be getting trained in the latest techniques?

One question was left off the list in the previous section and it is not one that the supervisor can answer. Having collated all the information you can – does the project excite you? Our advice is to go with the projects that have the most ticks and, critically, where they have a large tick in the happy/excited box. There is little point in opting to do a project that, despite having everything in place, fails to stimulate you. The difference between a good project outcome and a mediocre one is enthusiasm. This can drive you to overcome the hurdles that can appear in the project and lead to new avenues to explore. Without it, the project can quickly become like a laboratory practical where it is simply a case of putting in the hours.

1.9 PROJECT ALLOCATION

The mechanism by which projects are allocated will vary from one university to another. In some cases, one academic has the unenviable role of trying to juggle everyone's wishes. In other cases, it is done directly by the project supervisor on a first come first served basis. If it is the latter, then it is obvious that you need to be quick, but that does not necessarily mean sacrificing the background check. The first thing to do is to schedule a meeting – this will buy you some time to scour the internet. At the very least, use the Meeting Checklist and, where possible, take notes.

There will always be a possibility that you will not get your first choice. There is nothing that can be done about it and while disappointment in such circumstances is inevitable, if you have followed the instructions thus far, then your second or third choices should still be pretty good alternatives. In either case, you know exactly what you are getting, what is expected of you and what you should achieve at the end. Compare that with simply jotting down any project or running at the very last moment to try to locate a project that is still available. There is nothing worse than having to spend months doing something you dislike. The journey now starts and the following pages will help you ride out the storms.

CHAPTER 2

Managing the Project

2.0 BEFORE YOU DO ANYTHING!

Research Project Success: The Essential Guide for Science and Engineering Students
Cliodhna McCormac, James Davis, Pagona Papakonstantinou and Neil I Ward
© Cliodhna McCormac, James Davis, Pagona Papakonstantinou and Neil I Ward 2012
Published by the Royal Society of Chemistry, www.rsc.org

Once the project allocations have been announced – the very first thing you need to do is buy a hard bound notebook with a decent number of pages. A near universal failing of students is a reluctance to keep an account of the work they have done and too often they rely on scribbles jotted down on spiral bound notebooks, post-it notes or, worse still, vague recollections from memory. It is common to hear things like "well the weight was about….I think…". The latter is often sufficient justification for a supervisor to bang their heads off a wall in frustration. It is vital that you keep accurate notes. The lab book is your record of what you have done so that you and others can refer back to it. It doesn't need to be a pristine piece of prose, but it should be detailed. It must contain your observations and records of measurements and it should also include your musings on what you think has happened. The latter is often ignored in the rush to simply follow the recipe that the supervisor has given. You must remember that you are expected to think independently and to critically assess the work that you are doing. Was the outcome what you expected? Did it conform to theory or was it a complete disaster? If it was a disaster, then what went wrong? It is a poor response to simply shuffle back to your project supervisor and report that it didn't work and ask "what will I do now?" You should document everything that you think is of value – and that includes your thoughts on the experiment itself and not merely the so-called hard variables like weights, measures and signal responses.

You may be tempted to keep only those records that relate to experiments that worked – after all, these are things that are important…aren't they? Well, no is the answer. Assessing what went wrong with the disastrous experiment is equally important and it is only by keeping accurate notes that you can go back and perform the necessary forensic deduction as to why it went wrong – and thereby endeavour to correct it in forthcoming experiments. Ideally, a record should be kept as the experiment is planned and executed, in addition to post-event critiques. It is no use simply jotting down the results after the event as memory is far from a reliable witness when dealing with minutiae.

In committing your observations and musings to paper, you have an accessible record of what was done and what happened and it is from this that you assess the progress of the project as you go. It is also an invaluable aide-mémoire in subsequent discussions with your supervisor or research team and also when writing up the work formally. The notebook should therefore be regarded as one of your treasured possessions. There is an increasing trend towards electronic notebooks and digital records, but at present, the humble sheet of paper is by far the more flexible in terms of jotting, sketching and doodling. The increasing accessibility of software may well change this in later years, however, there is one issue that will be forever present – lost data. Most instruments in the laboratory will be computerised and therefore you can expect the data recorded on them to be stored and regularly backed up. The same is not always true of the portable memory devices that you carry – can you remember the last time you made a backup of your USB memory stick? Worse, are you unfortunate enough to recall one malfunctioning or being lost? It is here that the humble, hard-bound, notebook proves its worth. Its bulk is one attribute that helps prevent its inadvertent loss though, admittedly, it does not stop it from being left behind in libraries, on buses, etc. It is, however, immune to corruption or being accidently overwritten. Nothing should ever be removed or erased from the book.

A factor that must also be considered is that a hard-bound notebook containing scribbles is going to be less important (or valuable) to someone than the latest USB storage device. As such, it has more chance of being found where it was lost and consequently there's a much greater likelihood of retrieval or, indeed, of it actually being returned to you – just remember to put your name, university and course code on it. If you are one of those who have a well-established relationship with your computer, then by all means put things down in electronic format to be held on a disc or in a virtual cloud, but please record the original data in the old-fashioned way.

2.1 FIRST STEPS

Remember: It is good to talk. Good communication
is key to success

Irrespective of the description used to describe the project and the level of enthusiasm displayed by your new supervisor, the immediate goal of the project is not to make a startling discovery that will lead to a Nobel Prize – your goal is to produce a document you may consider somewhat humble but that will assure you an excellent mark. It is important to remember that the project will carry a significant weighting with regards to your degree classification and therefore you want to make sure you can get as many marks as possible. This is your main concern. It is therefore vital that you focus on the dissertation from the very beginning rather than leaving it to the very end – as far too many students do. It can be very tempting to convince yourself that as the dissertation is not due to be handed in for many months – it can be largely ignored – for the moment at any rate. There will be many other things weighing on your mind such as lab reports and assignments and, after all, you need to get results first – don't you? This is a critical mistake. It will mean that you are left running to complete the write-up at the very last moment and probably at a time when so much else needs to be done. This will inevitably cut into time that should otherwise be dedicated to exam revision.

The worst thing you can do is submit a shoddy piece of work. Not only will you lose out on much needed marks, but you will end up shooting yourself in the foot when looking for a job or postgraduate course. A potential employer will invariably ask about your project during the interview (Box 2.1). This question is meant to be the ice-breaker. It is intended to put the candidate at ease, by allowing them to talk about something they should be comfortable with. It is also the perfect opportunity to bring out the project dissertation and use it as a means of demonstrating to your prospective employer those all important transferrable skills of presentation and communication. It effectively serves as a showcase that you can use to highlight the quality of your work. That, of course, only holds if it is well done.

BOX 2.1

In contrast to the multitude of lab reports you have completed the project dissertation has a life of its own and it can either be an invaluable aid or, if poorly constructed, a skeleton which, at best, needs to be hidden, but may actually come back to haunt you – especially if the interviewer expects to see it or asks for it directly.

It is therefore vital that you set out a plan so that you can complete the dissertation as you go rather than leave it to a mad rush the week before the deadline. Once the projects have been allocated, the best course of action is to contact your new supervisor and arrange a meeting. This should be done as soon as possible. It may be that the project allocations are announced well in advance of their starting date. This is actually the ideal situation as it gives you more time to plan. Some departments allocate their projects at the end of the second year precisely for this reason – giving you all summer to think about the work! No matter when the project alloca-tions are announced – you must make the effort to contact your supervisor – it is unlikely that they will chase you. This is where the differences between project and conventional lab sessions start to appear. You are the person who must drive the project – the acquisition and the analysis of the results are very much your responsibility with the supervisor providing the guidance when called upon. It is up to you to control how the project is conducted. Communication between you and your supervisor is key.

2.2   ARRANGE YOUR FIRST PROJECT MEETING

The first 'Project Meeting' should seek to set out how the project is going to be conducted and from that you should then be able to put the foundations of the dissertation together. As the project progresses, you can slowly build it up. This meeting should flesh out the project that you are about to embark upon and ensures that you are heading in the right direction. It is at this stage that you should be prepared to take detailed notes on what the aim is, the final outcome and, crucially, what the various steps are. The latter are important as you can use these as a guide when assessing your progress. Various terms can be used to describe the intermediate steps and can include: phases, stages or work packages. These phases can often be regarded as the broad objectives of the project. The aim is to get to the end. The objectives are the steps in between.

To get an insight into what each stage of work will entail, you really need to have a conversation with your supervisor. There is a danger, however, that during the first meeting you simply sit and nod your head in agreement. HUGE MISTAKE! You should be taking notes and asking questions at this stage. If you are not writing furiously at this meeting, then you are not going about it the correct way. You need to listen to what the supervisor is saying – you must listen intently and then be bold and ask them to explain in more detail if you don't understand the approach they are taking/suggesting. This is your project and it is therefore up to you to ensure that you amass all the information you can. You still have to do the literature review and the information obtained from the supervisor during the first meeting will save you an incredible amount of time – but you must ask (Box 2.2). You must chase the answers.

BOX 2.2

The assessment of the projects will vary from one university to another, but they will be similar in format. In general, the mark scheme will be split into two: the practical work and the dissertation. In terms of the practical work, there will be a significant proportion of marks allocated relative to the level of enthusiasm exhibited by the student. Did they contribute meaningfully to the direction of the project or did they simply follow instructions? It is obvious that the former will command the most marks while the latter will barely get you a pass. It is therefore vital that you are seen to be fully participating and providing critical evaluation of where things have gone wrong along with suggestions as to how to proceed.

Deadlines should be put down for the completion of each stage. These will be tentative and may well change, but it is important to have something to work towards in the initial phases. Most important, however, is the need to clarify the deadlines for the submission of assessed work – be it preliminary literature reviews, the dissertation itself, posters or presentations. These are immovable and should be in your mind from the start as these will count towards the final mark.

2.3   PROJECT OBJECTIVES AND MILESTONES

Use milestones to chart your progress

All too often students simply rely on being told what to do. They may have an idea of the grand plan, but the route from beginning to end can be very vague. The problems arise when trying to write it all up when you are left with a collection of experiments, but have no idea as to how they all fit together. The project objectives are the intermediate steps necessary to get you from start to finish. It is important that you sit down and break the project up into stages – these need not be a linear progression, but could be conducted in parallel. What is important is that there are well-defined points where you can evaluate the progress of the project. If the project involves making something, then the objectives could be the intermediate stages in the manufacture of the material, drug or device, etc. Let's take a more specific example. Consider the project outlined in Figure 2.1.

The project objectives could be:

- construct copper electrode sensor;
- optimise operating characteristics of the sensor;
- assess response to nitrate;
- evaluate accuracy of the sensor.

Note that the above headings are very broad and while it is possible to break them down further or expand the list, it is normally sufficient to specify only the general themes of the stages. Each stage should be allocated a certain amount of time to complete and thus each milestone represents a time marker that not only indicates the direction in which you are going, but how far you have come and need to go. This is where you will need the guidance

---

Development of a sensor for assessing river water pollution

Nitrate is a common pollutant in surface waters and there is a pressing need to develop new diagnostic techniques that can facilitate its measurement with a high degree of accuracy. The project would investigate the application of a nano-structured copper electrode to the electrochemical reduction of nitrate and assess its suitability for use in decentralised sensing applications.

Figure 2.1 A typical project description.

of your supervisor as they will provide you with a reasonable guesstimate of how much effort each stage (or work package) will take. The more difficult aspects will obviously take longer.

Milestones are generally associated with each stage of the project but not necessarily to signify completion – rather they indicate an important point in time where the outcomes of the work are assessed so that a decision can be made as to whether to progress to the next stage. Once the milestones have been decided upon then it is customary to construct a Gantt chart (Figure 2.2) to highlight how you expect the project to progress. These charts can be prepared in a spreadsheet or drawing package. The use of dedicated project management software is rarely required for most undergraduate projects.

The stages need not be conducted in series and there can often be a parallel development pathway or partial overlap – especially where a previous design/product or process is adjusted as a consequence of the results obtained from the subsequent stage. This is a form of iterative optimisation.

Thus, in the example detailed in Figure 2.2, the student would be expected to produce some copper-modified electrodes – this would initially involve familiarisation with the core techniques required to produce them. The first milestone would signify an important point in the work package where the decision should be made as to which

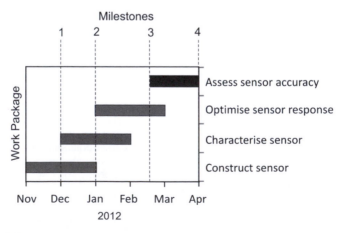

Figure 2.2   Gantt chart highlighting the project milestones.

ones are used in the subsequent testing stage. There is a degree of overlap between work package one and two as those electrodes that are deemed the best (work package 1) are tested (work package 2) and then further modified/optimised (revisiting work package 1).

2.4 ESTABLISH A REPORTING PROCEDURE

The next thing to do is to put in place a regular time/day when you can meet your supervisor. Regular meetings with your supervisor are essential to ensure that you are not left to drift off in the wrong direction and to ensure that the work progresses. It enables both parties to sit down and evaluate whether something has worked – or if it hasn't – whether to try again or to seek an alternative route. There is little point spending weeks in the lab with an experiment that fails to work, but that you stubbornly repeat over and over again in the hope that it will eventually succeed. It may be that you tweak the experiment each time in an attempt to get it to work. It would be a much better approach if you were to seek the advice of your supervisor – explaining what you have done so far, where you think it might be going wrong and what you were thinking of doing to overcome the hurdle. Note that the emphasis is on you. You must critically appraise the situation and then seek counsel before moving on. The worst thing you can do is simply report to the supervisor: "It doesn't work. What will I do now?" Remember that you have to take responsibility for the project and therefore you must invest the time in trying to understand why it didn't work out as planned. The significance of this approach is that you will be rewarded in the subsequent assessment of your practical work.

2.5 SUMMARY/KEY POINTS

- ☑ Buy a good hard-bound notebook.
- ☑ Arrange a meeting with your supervisor.
- ☑ Ask your supervisor to explain what is involved AND take notes!
- ☑ Clarify the project objectives.

☑ Identify the milestones – points at which the project progress is assessed.
☑ Clarify deadlines.
☑ Prepare a Gantt chart highlighting the relation between work stages.
☑ Establish a routine reporting procedure where the project is discussed in detail.
☑ Manage each stage through carefully assessing progress.

CHAPTER 3

Searching the Literature

3.0 WHAT IS A LITERATURE REVIEW?

Research the background to the project

The literature review is an integral part of the research project as it allows you to get a feel for the subject area and to get a grasp of the

Research Project Success: The Essential Guide for Science and Engineering Students
Cliodhna McCormac, James Davis, Pagona Papakonstantinou and Neil I Ward
© Cliodhna McCormac, James Davis, Pagona Papakonstantinou and Neil I Ward 2012
Published by the Royal Society of Chemistry, www.rsc.org

significance of the topic in a real-world context. It should also give you a framework onto which you can build a robust understanding of the core concepts and the work others have done in the past. However, it is more than just a mammoth reading session as you will be required to produce a concise summary of the research that relates to the project you are about to undertake. It is quite common for the literature review to be assessed separately from the project work and it can fall under the banner of Research Methods or a similarly titled module. There are merits to this format as it forces you to think about the dissertation from an early stage and should, ideally, improve the quality of the practical work as you will know more about the theory, the reasons you are heading in a particular direction and to assess how your results fare in comparison to other groups.

The literature review is more than a simple essay as, irrespective of whether or not it is marked separately, it will still be included within the final project dissertation. It will therefore form the bulk of the first chapter and it is arguably the most important part. Why? The introduction and background will be the first thing the examiner reads and it is crucial that you outline the significance of the work you have done and why you are doing it. The examiner may not be an expert in the area in which your project lies, therefore the first chapter must convey to them a feeling of what it is you are trying to achieve. In essence, you need to spark their curiosity and, in doing so, there is a good chance you will put them in a better frame of mind. Remember, examiners are human and there can be few things that are more depressing to them than having to wade through a pile of poorly written dissertations. The appearance of a well-presented thesis that attempts to carry the reader through the journey is often a boon to an examiner and more likely to be rewarded – especially when it stands out like a beacon in comparison to the others. The cue therefore is to ensure that you engage the reader at the outset. Ideally, your writing should make them want to read the rest of the dissertation or, at the very least, ensure that they don't let out a sigh after reading the first page.

The review itself is more than simply saying that this research group did this and another group did that. You must demonstrate to the examiner that you have not only read about the subject, but that you understand it and are able to critically appraise the work produced by others. The key concept here is 'critically appraise'. You have to be able to recognise the merits and limitations of different approaches and present them in a coherent, professional manner. It is worthwhile re-iterating that when discussing the results you have obtained in the practical phase of the project you need to be able to put them in context and hence compare them with the results others have achieved. Failure to do so could cost you dearly when the examiner comes to allocating marks. Simply presenting your results without duly assessing their value will gain few marks.

3.1 BEGINNING THE SEARCH

That wasn't quite what I meant when I said
you had to dig deeper

You should already have started your search of the background material when you were gleaning some details to help you in the project selection process (Chapter 1). Now you have to dig a lot deeper. This process should be started and, ideally, completed before you enter the laboratory. The temptation to get down to the actual practical work may be overwhelming, but such an approach is foolish in the extreme. The initial steps must be to get an understanding of the background first so that you fully appreciate what it is that you are going to do and why. It is acceptable to enter the laboratory to receive training on various aspects of the techniques you will be using and sometimes this may be necessary to work in with the schedule of the supervisor or technicians, but the background/literature review must not be put off at the expense of pottering about in the lab.

3.1.1   Textbooks

The most obvious question is where to look – books, journals or Wikipedia? Is one source more likely to yield a greater reward than another? Unfortunately, that is not a question that can be answered simply. The library catalogue should be your first stop. A word of caution: don't expect to find a book on the shelves that corresponds to the title of your project. All too often students complain that they cannot find reference material. This can infuriate a project supervisor when they subsequently find out that the keywords have been those of the project title and nothing else. It is likely that aspects of the project may be held within textbooks with more general titles. The question you have to ask yourself is what subject area does it fall under? You then find the shelf mark and peruse the books looking for a chapter or a section heading. Sometimes you may be lucky and there may be a whole book devoted to your project area. Although textbooks can be a good place to start, remember that the information held within them is likely to be out of date by the time the book is actually published. They can be invaluable for ascertaining the foundations of the subject but they are not a reliable gauge as to recent developments.

3.1.2   Journals

Journal articles will be the second port of call on the expedition to find out about the subject. In days of yore – you would have happily skipped from the books to the periodicals section of the library – but nowadays, you will find few printed editions. This is an increasing occurrence in modern libraries, not because there is a decreasing number of journals – far from it – it is simply that the majority of journals are now accessible as online editions. There is seldom the need to enter the library other than to check on the book list. There are numerous reasons for this – economics being a major one – but there is no doubting the flexibility that online searches provide. The demise of the printed edition is a shame in some respects – beyond the mere nostalgia of a dying age – as the rifling through the hard copy could often lead to unexpected finds that could often stimulate thoughts about how the project could be extended in different ways – not to mention articles that were just simply interesting in their own right. There is much less tendency to browse beyond the results of a keyword search. If you get the chance and your library still has hard copies – have a leaf through the hard copy of a journal. The authors wager that you will find something within its pages that will catch your imagination – even if it is irrelevant to your current project.

Most libraries have been re-branded as 'Learning Resource Centres' to reflect the advances in technology and that they are not simply a repository for books, but are now the central hub through which you can gain access to electronic resources. The latter will include a large number of journals which can be searched online. As part of your undergraduate training, you should have received lectures or tutorials on how to use the library. At the time, this may have seemed an irrelevance, but it is now that your ability to negotiate the various databases becomes crucial. As the majority of literature searching is online, you need to know which resources are available and how to access them. In most cases, universities provide off-site/off-campus access to these learning resources that avoids the need to trek to the library itself. There are generally three broad options for online searching:

1. Publisher portals (i.e. Science Direct, Biomed Central);
2. Web of Knowledge/Web of Science (WOK/WOS);
3. 'Wiki' and Websites.

Publisher portals can be useful, but it is important to note that you are restricting your keyword search to the journals promoted by that particular publisher. Take two examples: Science Direct and the Royal Society of Chemistry's Portal. Both allow keyword searching across a number of journals, but neither portal will bring up results from the others' repository. A frequent failing of students is to restrict themselves to Science Direct. It is true that it holds a tremendous number of titles and it does possess a highly versatile search engine, however, any search is restricted only to Elsevier titles and there is a wealth of information available from other sources. Does this mean that you have to visit every publisher website and repeat the search over and over? Fortunately the answer is no.

3.1.3   Web of Science

The Web of Knowledge/Web of Science is an abstracting service that essentially takes the bibliographic details and abstracts from articles published throughout the world and stores them in a central electronic archive. It is possible to perform a keyword search that has the potential to pick up articles independent of publisher. There is a caveat, however, it is not universal and there is still a large number of journals that are not registered with the system and are thus, effectively, invisible to its search engine. It can still be useful to check individual publisher sites for journals that could fall under the radar – at least as a secondary measure. Nevertheless, the catalogue of journal abstracts held by WOK/WOS is highly impressive and likely to satisfy the majority of undergraduate and postgraduate research studies.

The WOK/WOS search engine allows you to search by keyword, author, address, etc. and offers various options for conducting more advanced searches – such as combining previous results. The main outcome will be a list of journal titles that meet the search criteria you specified. The system will present the full bibliographic information and a hyperlink to the article abstract (if there is one). The steps involved are highlighted in Figure 3.1.

It will often be necessary to refine your search – especially if you have used fairly common words. Consider the nitrate project example outlined in Chapter 2. The basic process is highlighted in Figure 3.2 where the successive addition of keywords is used to whittle down the search results to those that are of direct interest to the project. Entering 'nitrate' as the keyword will generate over a quarter of a million hits – most will be irrelevant. It is unlikely that you will be able to scan through these to find those that are directly related to your particular project. Adding more keywords to the original search string such as 'copper' and 'det*' will bring the number of hits down considerably but it still presents a reading list that is unmanageable.

In most cases the search engine will accept wildcard operators such as '*' – which allow partial word matching. The example used

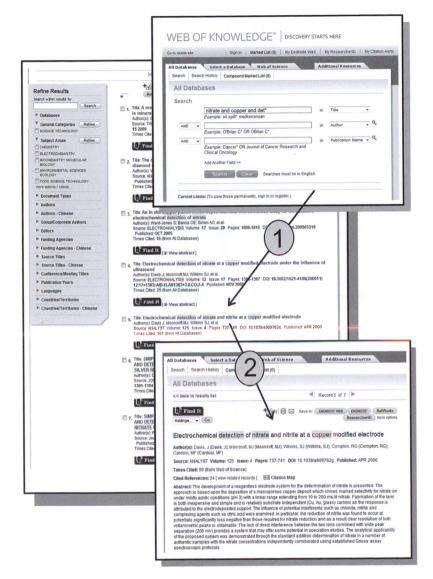

Figure 3.1 Search for articles in WOK/WOS.

in Figure 3.2 used 'det*' which will pick up 'detection', 'detecting' and 'determination' all of which are relevant. An important note is that when performing a broad search, through the topic for example, the scan will simply pick up instances of the sought after

Figure 3.2   Keyword searching in WOK/WOS.

words irrespective of how they are arranged. Thus, it is very likely that there will still be a large number of hits that are irrelevant. If you are looking for a specific phrase rather than a word then encase the key phrase within quotation marks. For example: using 'copper' and 'electrode' will generate a large number of hits as the search engine registers hits for each record that has those words anywhere in the abstract or title. In contrast, using the term 'copper electrode' within the quotation marks will restrict the search to those articles that involve the use of a copper electrode.

Initially the search string was applied to the topic – this encompasses the title, keywords and the abstract. This is a first stage and can then be refined. When the keywords have been exhausted and a large number of hits remain – the next step is to change the target and examine only those articles that have the search words in their title. In doing so, you are looking for those articles that focus on the measurement of nitrate in which copper plays a central role. This should be the final step as it narrows the search remit greatly and, in all likelihood, will exclude a fair portion of articles that would be of relevance to your project despite not having the keyword in the title. In general, it is better to try to refine the search through keywords, but if all else fails then title search should be considered. Note that it is better to start from a high number of hits and work down, stepwise, rather than putting everything into the search string and restricting the search target to title. All too often students will attempt the latter and then report back to their supervisor that there are no articles on the subject. They have been far too specific. Start from the most general first and then work down to a number that is manageable.

The example used above involved the detection and measurement of nitrate in river water. It is important to consider the words that you enter as the search string (Box 3.1). First, decide on the nouns to be used – search the project description for proper nouns. In this case 'nitrate' and 'copper' are the ones to use. Including verbs within the search string can often present a dilemma for the would-be review writer as there can be a range of synonyms and their use will depend, at least in part, on the literary style of the author. Hence the word 'measurement' is not as common as 'determination' in journals. The two may mean the same but the former is the one you would use in everyday conversation, whereas the latter is the more formal equivalent. Entering 'measurement' in the search string could result in a large swathe of articles being overlooked. This is exemplified by repeating the process detailed in Figure 3.2, but substituting 'measurement' for 'det*'. The outcome is 62 hits before changing the search target to title (compared to 2088 arising from the use of 'det*'). Refining the search to target the title will result in zero hits! How do you know which synonym to use? It comes down to experience. As you scan through mounds of articles, you will begin to see recurring words and phrases. For the present case, however, the best approach is to try to restrict

yourself to nouns where possible or at least for the initial search
and then, if faced with a large number of hits, introduce the
appropriate verb thereafter giving consideration to repeating the
process for its synonyms.

BOX 3.1

It can also be an idea to add in the word 'review' to the keyword
search. It is likely that someone will have previously conducted a
literature review on the topic and therefore will have done some
of the legwork for you already.

3.1.4   Extracting the Article from the WOK/WOS System

One of the most important benefits of the WOS system is that it
provides the full bibliographic listing plus the abstract. The latter is
a vital component as it should contain all the important information that you are actually looking for. The abstract – by its nature –
summarises the main findings of the investigation and should
enable you to assess whether or not it is worth chasing down and
looking at the whole article.

   Acquiring the journal articles has also been greatly simplified by
the shift to electronic access and it is increasingly common to find
within the WOS listing a button that will direct you to view 'Full
Text'. There can, however, be a misconception that this means free
access – this is not always the case. While there are increasing
numbers of articles being published under free-access programmes –
the majority, at least at the time of writing this book, are not and
when redirected to the publisher's site – you will be asked if you wish
to purchase the article. All is not lost here as it is possible that your
institution subscribes to that journal. When you embark on a university course you are normally allocated an 'Athens' password –
this may not be given to you directly but will become electronically
ascribed to your student ID and library password. If you are conducting the search through your university's library portal – the
Athens 'passport' normally accompanies you when perusing the
various electronic resources. If your institution is a subscriber to a
particular journal then it should be a simple case of clicking on the
pdf icon to obtain a copy of the article.

When conducting the search off-campus and you are confronted with a publisher site demanding monies, it is worthwhile checking on the website to see if there is an 'Athens/Institutional login' button. This will temporarily redirect you back to the university's server for you to enter your details. It is important to note that few universities have the resources to subscribe to every journal and many will buy 'packages' from the publishers. These tend to consist of the most popular titles across the various science disciplines.

In addition to the 'Full Text' button, there will often be a secondary 'Find it' option. The latter will be integrated with your university's access subscriptions and will attempt to find the article by directing you to the appropriate publisher website. If the article you want to look at is in a relatively obscure title or a new journal, there is a strong possibility that there will be no electronic access to it. This gives rise to the question – do you really need it? This can be difficult to answer as you obviously won't know what it contains before you get it. The abstract can give you an indication. If there is no abstract then you need to think carefully. We will return to this situation in a moment when considering how to search articles for information. The quick answer is that if you are confident that you need it then you will need to request an Inter Library Loan (Box 3.2).

BOX 3.2

Interlibrary loans involve your university either borrowing a book from another library or obtaining a copy of the article. You need to ask your librarians what the procedure is for obtaining such a loan. It could be in either paper or electronic copy. In any case, you will need all the bibliographical details. Such loans can take several weeks to come through and there is a cost to the department – you will invariably need the signature/ approval of your project supervisor.

An alternative approach can be to contact the author of the paper (normally there will be an email address for the author with responsibility for the article somewhere on the front page) and ask to be sent a reprint. This can be worth trying, but the response will depend on the generosity of the author. Responding to such requests can be time consuming despite the wonder of email and you may find that your request goes unanswered – not through any premeditated intent, just simply as a consequence of being lost in an ever-increasing 'to do' list.

Once upon a time you would have to trek to the library, locate the journal within the bookshelves, dig out the article from the heavy volume and march to the photocopier. Now you can sit back in the comfort of your home and at the click of a button have the article delivered to your desktop. The downside is that you could easily be swamped with lots of articles – some relevant – some not so.

3.1.5   Patents

Most students neglect patents as possible reference sources. There is increasing pressure on academics to protect their research results – their intellectual property – and they will often submit their work in the form of patents before publishing in the peer-reviewed journals. The WOK/WOS system will not pick these up. Scifinder Scholar is similar to WOK/WOS, but will pick up both research articles and patents. Sadly, access to it is limited and most universities will have modest site licenses. Nevertheless, it is worthwhile enquiring to see if there is a computer on which it is installed. The alternative is to use commercial web-based search engines. A quick internet search will reveal a number of options, but there are too many to list here or describe. Most will provide keyword searching in the same way as that described for the WOK/WOS systems and will invariably provide abstracts of the patents.

3.1.6   Web Pages and Wikipedia

Internet search engines can be extremely useful for finding background information, but are web pages acceptable for inclusion within a literature review? There was a time when the answer would have been a resounding no. Your supervisor could be reliably predicted to go into a state of apoplexy at the mere mention of a web reference. There can, however, be a place for web reports and their inclusion is becoming almost unavoidable. Most organisations – whether government department, private enterprise or charity – are increasingly putting their information online and, in many cases, their pages can be an excellent and convenient reference source – especially for statistics. More often than not the web page will simply present a previously written report or a document file or spreadsheet that can be downloaded.

There is, however, a need to be highly cautious when using the web as a reference source and you need to exercise some critical judgement as to the nature of the material you are viewing. The primary difference between a journal article and a web page is that the former has been peer reviewed while the latter is not. Peer review is a safeguard that attempts to ensure that the content of the article is accurate and free from bias. That is not to say that the majority of web pages are not. Web content delivered by enthusiasts can be of equal value to any book or journal article when learning about the subject you are researching. In fact, they can often be far superior to a book where multimedia is used to exemplify core concepts.

Would you include such material in your review? No.

Why?

Part of your literature review is to provide a road map for other researchers to follow. What if the web page is not there in the future? Books and journals, whether in hard copy or e-format, have a fixed content and will be archived by the publisher for future review. A webpage by an enthusiast or company is transient. It can change or disappear on a whim. A case in point is the use of the WOK/WOS system where the design of the web portal is liable to change over time. It is likely that by the time this book goes to press, the screen shots shown in Figure 3.1 will be different – fortunately, the search process outlined above will still operate irrespective of how the front page is tweaked.

We suggested that books be the first port of call – we would hope they are – but, in reality, there is very good chance it will actually be Wikipedia. This resource could be almost considered as an intermediate between book and journal, but, in contrast to journal articles, Wikipedia can evolve to take into account new developments. It is a superb resource for quickly gaining the 'feel' for a topic and, quite often, there can be some useful references to support the content. Your supervisor may tell you that you shouldn't use Wikipedia – yet there are probably few who don't use it themselves. Hypocrisy? No. They are correct as the factors that make it such a superb resource are also its main weakness in terms of being acceptable for inclusion within a professional report. It is essentially a massive web repository whose content is maintained by enthusiasts. The content is increased and modified daily. It could be argued that it is subject to a form of peer review as there is an option for those browsing the articles to update or correct the material. So what is the issue with using it in the report? Well – it can change and the quality of the peer review can be variable.

Traditional peer review is ideally where the material is reviewed by a panel of experts (typically three) before being recommended for publication – there are no subsequent changes. Effective peer review requires input from specialists who are familiar with the field on which the material is based. It is conceivable that a page could be constructed on a particular research topic by a student. While the broad aims of a research project can have a potentially large audience – interest tends to fall off sharply as investigators focus on particular bits. In some respects, a research project is like a painting – people like to look at the picture but they have no interest in the intricacies of how the paint was prepared. Thus, an error in the web page could easily remain unnoticed for a large period of time but the consequences are that other students searching on the subject will then propagate the error. Will academics not pick up the error? Yes! When they mark your work and give you a low mark. Just as we mentioned that requests for reprints may fall victim to an academics expanding 'to do' list, so too may their inclination or intention to view or correct the web content on Wikipedia. We would recommend using such web sites only as a means of gaining an overview and occasionally for reinforcing your understanding of the subject.

3.1.7 Finding a Balance

The web-based sources could be included within the introductory section of the report alongside book references to highlight the fact that you have read the core theory that underpins the project. The bulk of the review, however, should be in the form of journal articles and preferably the most recent. If you are confronted with a large number of possibilities then focus on the most recent years. It is important that you do not forget to add in seminal papers that underpin the work. A field that has evolved over many years may still rely on work that was done decades before. How do you find such work? There will invariably be a reference to it in the more recent articles and if it is that important then it should be a recurring sight.

How many references do you need? As many as is appropriate. This seems like a glib answer, but each project is different and depends on the objectives agreed between you and your project supervisor. The worst thing to do is to base your literature review on one or two books. This will scream out to the examiner that you have not searched the literature at all. The second failing is to present a list of URLs. This will suggest that not only are you lazy, but that you don't know the relative value of different source materials. Aim for a few books, a few web-based reports and the bulk composed of journal articles (whether hard copy or e-journal).

3.2 FORENSIC REFERENCE HUNTING

Discover the thrill of detective work...follow the citation trails.

There is another option that is complementary to the keyword search approach described above. Ask your supervisor for some suggested reading material. All you need is one article and from that you can work backwards to find articles of interest as indicated in Figure 3.3. There is, however, a possible flaw – it

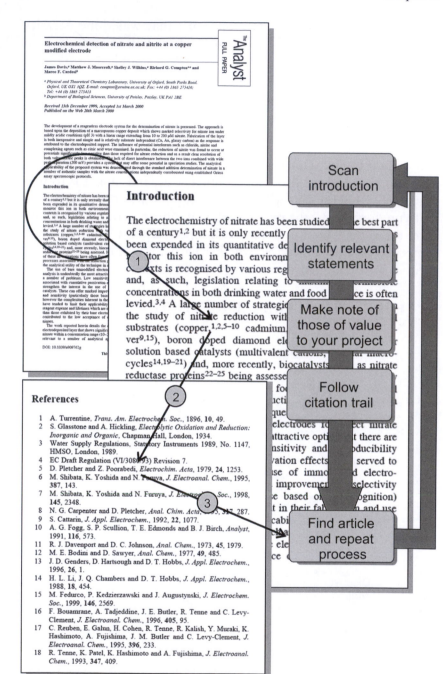

Figure 3.3 Following a citation trail.

depends on when the article was originally published. As with books, research articles can quickly become dated, especially if there is a lot of activity in the field of study. Even if the article is several years old, most of the articles will still be of relevance. If, however, the work is many years old then the chances are that things have moved on considerably. The remedy, however, is to find a more recent, preferably current, article. Use the keywords present in the article to locate the more recent article using WOK/WOS and then work back from that.

A word of caution: make sure you do the legwork and follow the citation trail. There can be a temptation to simply cut and paste the statements from a paper and add it into your review – along with the citation. Cutting and pasting falls well within the definition of plagiarism (discussed later in this chapter) so does the lifting of a citation without examining the original source. It is lazy and you will be caught out should you have to defend your thesis and it becomes clear that you haven't actually read the literature. Use the process outlined in Figure 3.3 to follow the trail and not as a means of creating an illusion that you have.

3.3 COMPILING THE LITERATURE REVIEW

Once you have accumulated the reference material, read through it and make appropriate crib notes. Now you are ready to start thinking about the form of the literature review. You need to work out a basic plan – as you would when beginning to write any essay. In this case the literature review must inform the reader of the following:

- an overview of the problem to be addressed;
- its significance – real-world context;
- critical assessment of previous studies – what was done and the outcomes;
- the issues that remain to be solved;
- the objectives of the proposed investigation.

It is wise to provide a brief explanation of what the overall aim is at the very outset so that the examiner is in no doubt as to the significance of the project and why you are interested in it.

Only highlight the aims at the beginning and not the objectives. The latter come at the end of report where you clarify what the issues/weaknesses are with the previous studies and then outline how your project will attempt to overcome at least some of them. Then the objectives make an appearance.

In cases where the literature review is assessed separately from the main dissertation, it may be necessary to expand the content to include a section that examines the various objectives in greater depth. It will be necessary to detail the techniques that you intend to use during the investigation – why you have chosen them and the results that you expect or hope to obtain. This is essentially where you are outlining the methodology behind your project. It may also be necessary to present a Gantt chart highlighting the milestones (Chapter 2). It is important that you clarify with your supervisor what is actually required for the assessed part of the literature review. If in doubt then put both sections in. When it comes to compiling the dissertation you should consider splitting the review and methodology components into two separate chapters.

3.3.1   The First Page

The first page is where you need to set the scene (Box 3.3) and is arguably the most important part of the dissertation from a reader's point of view as you must convey clearly the significance of what it is you are trying to do. You should have in your mind that there will be a readership of three – two examiners and your grandmother! The dissertation has to be correct scientifically, but it must also be written in a style that is readily understood and it must be in the third person (Box 3.4).

BOX 3.3

It is vital that the first page clearly outlines the following:

- the problem to be addressed;
- the significance of the problem to the wider world;
- the aims of the present project.

BOX 3.4

Style conventions:

Make sure that the dissertation is written in the third person. There should be no I, we, me or our within the text. Consider the following two examples:

1. "I found that the absorbance was linear with concentration. . . ."
2. "The absorbance was found to be linear with concentration up to. . ."

While both are correct only example two is appropriate for inclusion within a dissertation.

Your Grandmother might struggle a bit with the science but the introduction itself should give her the basic idea of what it is you are setting out to investigate and why! In essence, the introduction is effectively your sales pitch. If the reader is left confused then it casts the remainder of the dissertation in a bad light.

You need to support your statements with reference to legislation or statistics. In the case of the nitrate project then consider the following:

- What is the prevalence of nitrate pollution in the UK?
- Why does it arise?
- What are the factors that mediate the process?
- How much does it cost to treat?
- Is there any legislation to regulate nitrate discharges?

It is imperative that you make bold statements to lend impact to your project – but you must do so secure in the knowledge that you have a 'hard' reference source to support them (Box 3.5).

BOX 3.5

Most mark schemes for a literature review will be split into aims, background, scientific content, references and presentation. The majority of the marks will be for the scientific content with more modest allocations for the remainder. Nevertheless, these small pockets of marks are easily attainable, providing you put in the effort.

It is no good simply stating that nitrate pollution is bad and will kill fish through eutrophification. Reference styles are discussed in more detail in Chapter 4 but, for now, it is only important to recognise that there must be some form of written evidence that you can direct the reader to in order to support your statement and that it is not simply a glib generalisation or a perception held by you.

It is vital that you specify what the aims are upfront and clarify why it is important to conduct the project. The last thing you want is for the reader to have to plough through your literature review yet fail to appreciate what exactly your interest is. Some people will put the 'Aims' under a distinct heading. That is one option, an alternative approach is to set the scene in the first paragraph – providing a concise summary of what the problem is and then specify the aims of the project. Formats are, however, a bit like fashion – one style of coat may thrill one person but not another – what is important is that irrespective of the cut, the garment possesses some critical components – like fasteners – whether zips, buttons or velcro. When you construct your literature review you are free to create your own style or follow the current trends but what we aim to do here is make sure you include the critical components necessary for a literature review and indicate roughly where they should appear in the text. Their exact position is up to you. All of this should, ideally, be condensed into a single page and should be the first thing that the examiner reads. The fine detail of what you have pitched comes later.

3.3.2 Project Background

This is where you set the scene in more detail by providing an explanation of how the problem arises. Taking the nitrate project as an example then you would outline the sources through which nitrate finds its way into rivers and the various biogeochemical mechanisms that influence it. The material required for this section is likely to be found in books as it represents the foundations of the system that you are about to investigate. If the project were looking at developing a drug for combatting a disease then the background would be the factors that lead to the disease and the effects they have on the patient and on society as a whole. Should the project involve examining the properties of a new steel material then the background could focus on existing materials; how they are produced and their limitations. Computational projects could look at existing models and highlight how they are applied to real systems and again highlight their limitations.

3.3.3 The Literature Review

This section is where you need to focus on current approaches to the problems identified in the introduction and in the background section. This part should be formed largely from journal articles and ideally focused on research within the last five years. There are obvious exceptions such as historical/seminal papers that have formed the foundations of particular lines of enquiry. When collating your literature database you should have made crib notes – effectively extracting the highlights from each paper – what were the advantages and limitations of each. This information should be readily available within the abstract (effectively a summary of the main outcomes), the introduction (justification of why the specified approach is needed) and the conclusion section (usually highlighting the advantages of the approach detailed). Does this mean that you ignore the middle part? No. You should skim through the middle part to get a feel for the data that was obtained and the range of experiments that were conducted – you will not get that from the abstract. Why bother when you can get the main points from the abstract?

Skimming through the experimental and results sections of a paper may, at least initially, be a bit daunting, however, it will allow you to develop an appreciation of the experiments required to form the basis of a coherent scientific study in your particular area. You will find that most journal articles will have a similar structure in terms of the techniques used and this is what you need to carry in your mind for use later on when conducting the practical component of your project.

Sadly, there is an increasing temptation to simply rely upon the abstract to extract information. The danger here is that you are missing out on the nuances of the experiments conducted and therefore all that you can produce from the literature review is a superficial overview. You need to demonstrate that you understand the significance of recent investigations and be able to highlight the advantages and limitations. You will be able to glean some of this from the introduction section but remember that you are relying on the author's judgement. The introduction, as indicated earlier, is a sales pitch giving you the opportunity to outline why an investigation was conducted and why you should read it. In all likelihood, another paper will come along and say it is better and will set out its sales pitch. It is through taking the time to skim through the results that you will build up your own opinion of what the core advantages and limitations are and thus why your investigation is needed.

So how should you structure the review component? There could be two subsections – existing/established and research approaches. Returning to our nitrate project as an example, then the first subsection would document existing commercial products or established/accredited products, technologies or techniques. You would briefly explain how they work and what their respective advantages and limitations are. The second subsection should be larger and would focus on new developments that are documented with the research literature.

Should you write a bit on each paper you have found? That would be one way, but it will be a pain to write and even more so to read as it will tend to be fairly repetitive. You may find that a lot of the papers collected will be similar in theme – sometimes there will be only incremental changes in procedure. A better approach is to construct a table and summarise the main features of each (Box 3.6).

BOX 3.6

What are the 'main features' of an investigation? Projects will vary, but the reasons for pursuing them usually come down to a number of core factors:

- cost/viability;
- accessibility/complexity;
- time;
- effectiveness/efficiency;
- safety;
- lifespan/stability;
- strength/flexibility;
- size – nano/micro;
- accuracy;
- sensitivity and range;
- statistical robustness.

Where you have a lot of articles that look and feel similar, group them together and only refer to the most recent. In the case of the nitrate project, the table could include a comparison of the different techniques that are used, the samples they have been applied to and how effective they are. Two approaches to tabulating data are shown in Figure 3.4. The first table highlights the lazy approach and the second is where you demonstrate to the examiner that you

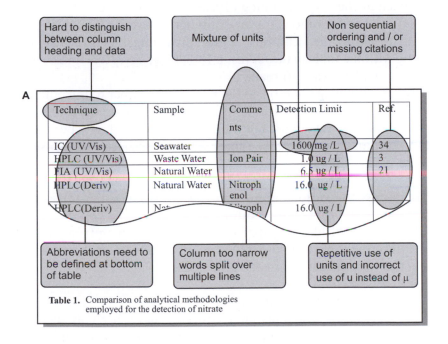

A

| Hard to distinguish between column heading and data | Mixture of units | Non sequential ordering and / or missing citations |

Table 1. Comparison of analytical methodologies employed for the detection of nitrate

Technique	Sample	Comments	Detection Limit	Ref.
IC (UV/Vis)	Seawater		1600 mg /L	34
HPLC (UV/Vis)	Waste Water	Ion Pair	1.0 ug / L	3
FIA (UV/Vis)	Natural Water		6.5 ug / L	21
HPLC (Deriv)	Natural Water	Nitrophenol	16.0 ug / L	
HPLC (Deriv)	Nat...	...itroph	16.0 ug / L	

- Abbreviations need to be defined at bottom of table
- Column too narrow words split over multiple lines
- Repetitive use of units and incorrect use of u instead of μ

B

Technique	Sample	Comments	LOD / ug L	Ref.
IC (UV/Vis)	Seawater		1.6	12
HPLC (UV/Vis)	Waste Water	Ion Pair	1.0	13
FIA (UV/Vis)	Natural Water		6.5	14
UV/Vis (Deriv)	Natural Water	Nitrophenol	0.10	15
ISE	Drainage Water		20.0	16
Voltammetry	Drinking Water	Cu Plated	5.0	17
Voltammetry	River Water	NR/Mediator	12.2	18

Where: LOD = Limit of Detection; Deriv=Derivative;IC = Ion Chromatography; HPLC = High Performance Liquid Chromatography; FIA = Flow Injection Analysis; GC = Gas Chromatography; UV/Vis = ultra-violet / visible spectrometry; CZE = Capillary Zone Electrophoresis; ISE = Ion Selective Electrode; NR = Nitrate Reductase.

Table 1. Comparison of analytical methodologies employed for the detection of nitrate

Figure 3.4 Common mistakes when constructing a table.

know your stuff, are willing to take time to present it effectively and that you want the marks! It is important that you make the effort as-remember – this document could be a useful showcase to woo a potential employer.

It is not always possible to present the table entries in terms of sequential citations, but try, where possible, to order them accordingly. You may wonder why the abbreviations are documented at the foot of the table. A list of abbreviations will be included at the very beginning of the dissertation (See Chapter 6), but it can prove arduous for an examiner to have to flick back and forth to check the definition. Thus, providing the truncated list along with the table makes it just a little bit easier. Remember, you want to make it easy to read.

Before throwing the table at the reader, remember to introduce it in the text first! It is advisable to have a small paragraph to summarise the contents (with relevant citations) and then where they can be found (i.e. Table 1). There are two scenarios to consider when using the table:

1. discuss the various approaches and introduce the table as a summary of the main features; or
2. outline the different approaches being taken upfront, introduce the table to compare the main features and then discuss the various merits of each approach.

An example of the latter is given below:

"A variety of instrumental techniques have been applied to nitrate detection but the majority employ either spectroscopic [12–15] or electrochemical detection [16–18]. The main features of each approach are compared in **Table 1**. The spectroscopic detection techniques fall within two categories: direct detection of the nitrate anion [12–14] and those where some form of derivatisation is required [15]."

Note that in the introductory text above 'Table 1' has been highlighted in bold – this is a minor presentation tweak and serves to catch the eye of the examiner. There is nothing worse than finding a table or figure later on in the dissertation and then having to scan back through to check where it was discussed. The bold highlight makes this process easier. Also note that there is no title to the table – instead there is a legend at the bottom. This is another stylistic tweak, but one that is regularly adopted within professional reports. Likewise, we would recommend that there is no title description placed on figures. Tables, figures and schemes, etc. should simply be given a legend rather than a title.

What about figures – should they be included? There is a saying that "a picture is worth a thousand words". This is only true if the figure is relevant and aids understanding. The inclusion of frivolous figures for the purpose of bulking out a thesis will not increase your mark tally. Flow charts to highlight or summarise processes can be very helpful as can reaction mechanisms, biochemical pathways or signal responses. If referring to chemicals in any detail then, where possible, provide the molecular structure. Obvious exceptions will be where enzymes, nucleic acids or other macromolecular species are involved. Make sure that such figures are properly labelled and don't simply appear from nowhere and are left floating on the page. It is also vital that you cite the source from which the figure was taken – even if you have modified it or made a version of it from a drawing package. In the latter case, it is necessary to state that your figure was "adapted from…" whatever source. The internet will give you access to a vast array of sketches, schematics, structures and molecular pathways – it is easy to copy and paste, but you must be judicious when including them in your dissertation and always cite the source from which it was acquired. Failure to do so falls under the remit of plagiarism.

3.3.4 How to Avoid Plagiarising Other People's Work

Plagiarism is just about the most serious crime you can commit in the academic world and is generally defined as attempting to take the work conducted by someone else – whether text, figure, table or their ideas/hypothesis, etc., and pass it off as your own. So what do you put in your literature review? You need to present your understanding of the work in your words and not sentences or paragraphs that have been directly copied. You must also acknowledge the work that others have done by citing them in the text and including the specific details of the source of your information in the bibliography section at the end of the thesis. So, are you allowed to reword other people statements? Yes….but you must cite the work that has led you to include this statement in your dissertation. The sentence structure should be your own.

Avoid superficial adjustments courtesy of a thesaurus where one word is simply swapped for another – such as 'detection' for 'analysis' or the simple reordering of the sentences within a paragraph.

It is acceptable to directly reproduce statements as quotes – providing of course that you cite the work. There is a caveat, however, in that quotations should be for pithy statements that have a distinct significance and should not simply be paragraphs that have been lifted from the web and encased within quotation marks. There should no screeds of text – copied and pasted into a dissertation, encased within quotations and cited. Technically, this is not plagiarism, but it is incredibly lazy and guaranteed to infuriate both supervisor and examiner. It indicates that you couldn't be bothered to expend the time or effort to construct your view on the subject. It is also guaranteed to score very low when the marks are attributed.

Make sure that when using images from the web the source is clearly stated. If copying an image from a printed journal, ideally you should get permission to reproduce the work in your dissertation. It can do no harm to contact the principal author and explain the purpose for which you want to use their image.

Make sure you consider the following checklist:

- When perusing the textbooks did you note down specific sentences?
- Have you copied and pasted anything (text or image) from e-books, e-journals or web pages?
- Did you remember to put direct quotes within quotation marks?
- When discussing ideas/concepts/hypotheses – have you cited the original sources?

What about sweeping statements that are widely known or often quoted but for which there is no apparent written or original source? Such information tends to fall under the banner of common knowledge. It assumes that the people working within a given field will have gained a foundation of key knowledge. An example would be: copper is a hard metal that conducts electricity. If you are working within a technology sector then such information is obvious and well established, will be known and accepted by a large number of people and as such, does not require a reference. If, however, the statement is more specific: copper ion can be used to catalyse the chemical reduction of nitrate – then the proportion of people knowing such facts will significantly diminish and therefore the source needs to be cited.

3.4 SUMMARY/KEY POINTS

- ☑ Ask your supervisor for background reading.
- ☑ Decide on the keywords for your search – concentrate on nouns rather than verbs.
- ☑ Explore the learning resources available within your institution.
- ☑ Search for appropriate textbooks for background theory.
- ☑ Learn to navigate the electronic databases – WOK/WOS, Science direct, etc.
- ☑ Collate a broad list of journal articles that may be relevant.
- ☑ Refine search to a manageable number (max. 50).
- ☑ Skim through the abstract and prioritise the articles.
- ☑ Summarise the content from each and, if appropriate, tabulate the main features.
- ☑ Construct a plan for the review.
- ☑ First page is your sales pitch – the significance of the project and the main aim.
- ☑ Background – provide more detail as to the problem to be addressed.
- ☑ Literature review –
     describe the existing practices/products/technologies;
     identify weaknesses that need to be addressed;
     outline what research is being done at present;
     critically compare and contrast the different investigations;
     summarise what the aim of your project is and why your
       project is needed in relation to current research.

The last point and the most important is that you cite your sources. The question that remains is how to do that and what form the citations take. That process is described in Chapter 4.

CHAPTER 4

Referencing Your Work

4.0 REFERENCE STYLES

Which reference road should you take?

There are a number of referencing styles and a quick scan through different journals will reveal that each can demand its own

Research Project Success: The Essential Guide for Science and Engineering Students
Cliodhna McCormac, James Davis, Pagona Papakonstantinou and Neil I Ward
© Cliodhna McCormac, James Davis, Pagona Papakonstantinou and Neil I Ward 2012
Published by the Royal Society of Chemistry, www.rsc.org

formatting style. So which should you use for your dissertation? The answer is simple – the one that the course leader says you should. It is likely that there will be departmental/course guidelines on the format your dissertation should take and within those there will invariably be a section on references. Two distinct forms tend to predominate: author–date and the numeric system. The latter is more common within physical science and engineering journals but the former, which has previously had a strong foothold within the arts and humanities sector, is now slowly becoming the standard for dissertations irrespective of discipline. At this point it is advisable to consult with your project supervisor as to which style is best for your particular work. Each approach has its merits and limitations and it is not possible to give a definitive answer as to which is best for your work. Both forms are, however, summarised here, for the most common type of reference sources that you will come across along with the core principles can be applied, with a bit of minor modification, to any style.

4.1 THE HARVARD REFERENCE SYSTEM

The Harvard reference system is by far the most common in universities and there are two components to its successful implementation: in-text citations and the corresponding bibliography. Both are required for your dissertation and it is usual for a small number of marks to be put aside to assess your ability to use and present then correctly. The in-text citation is effectively a curt signpost to the more extensive bibliography. The former is a marker to indicate to the reader that the statement/fact you have presented was previously considered by another researcher – whose full address can be found at the end of the dissertation. This is obviously not the home address of the author but rather the address where the source material can be found. The bibliography must give the reader sufficient details to find the original source quickly without having to do any guess work. The signpost itself is invariably the authors surname followed by the year of publication and, if appropriate, the page that you are specifically referring to. An example of an in-text citation is highlighted in Figure 4.1.

There are some rules associated with the presentation of citations and they usually revolve around the number of authors. If there are less than four authors, then it is customary to present all three

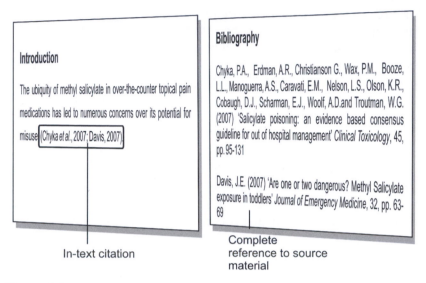

Figure 4.1 Harvard citation and bibliography formats.

names within the in-text citation. Four or more authors and the text becomes unwieldy and the in-text citation is shortened to the first author and followed by '*et al*.' Remember to italicise the latter and include the period after the 'al'. Typical examples of the in-text citation are shown in Table 4.1.

4.1.1 Journals and Newspaper Articles

There are numerous ways in which to cite the work of others and it will depend largely on your own writing style and the weight that you perceive a given reference to have. Usually you will bolster your arguments with work that underpins it and often the citation will come either part way through or at the end of a sentence and will be encased in parentheses. Occasionally, however, you may feel the need to specifically comment on a particular piece of work. In such cases, you directly acknowledge the authors name and only the year is placed in parentheses. In many ways, deciding what needs a citation, when and where, are the hardest parts of writing the literature review. The bibliography, in contrast, is perhaps the easier part to construct, but it requires a deft touch to ensure that it is correct. The Harvard style demands a particular format and its

Table 4.1 Harvard in-text citation styles.

Sources	Example in-text citation style	Comments
1	Materials such as boron-doped diamond (Jones, 2012), graphite (Hardy, 2010) and	When there are two sources attributed to the statement you are making then the author, date format remains the same but the two different sources are separated by a semi-colon.
2	Various forms of carbon have been used in the manufacture of electrodes (Jones, 2012; Hardy, 2010).	
2	Various forms of carbon have been used in the manufacture of electrodes (Jones *et al.*, 2012; Hardy, 2010).	
3	Previous investigations of composite fracturing (Jones *et. al*, 2010; Hardy, 2012; Smith and Fret, 2009) were focused on. . .	
2	The initial oxidation of graphite by permanganate showed little change in voltammetric resolution (Jones, 2010). Longer pre-treatment times were found to provide a more extensive potential window (Jones, 2012).	When citing multiple sources from the same author, two situations can arise – publications in different years or in the same year. The latter requires that the in-text citation of the two are distinguished by means of 'a', 'b' etc.
2	The initial oxidation of graphite by permanganate showed little change in voltammetric resolution (Jones, 2010a). Longer pre-treatment times were found to provide a more extensive potential window (Jones, 2010b).	
2	Initial studies of graphite oxidation (Jones, 2010; 2012) demonstrated that	Instead of repeating the first author name – a more succinct style can sometime be used: (Name, year of Source 1; Years of Source 2,. . .) Note that a semi-colon is again used to distinguish between the two different works.
2	Initial studies of graphite oxidation (Jones, 2010a; 2010b) demonstrated that	

conventions can be quite quirky until you get used to them and will depend on the type of material you wish to reference. The main approach for the referencing journals and similar articles is detailed in Table 4.2.

Table 4.2   Harvard formatting style for journals and newspaper articles.

Authors	*Bibliographic style for journals*	*Rules*
1	Jones, M. (2012) 'Laser ablation of Carbon', *Diamond and Related Materials*, 23, pp. 12–14	1. Surname goes first followed by a comma and then the initials. Two authors should be separated by an 'and' with multiple authors by a series of commas and last two names by 'and'.
2	Jones, M. and Hardy, D. (2012) 'Graphite Oxidation', *Journal or Molecular Catalysis*, 42, pp. 302–3	
>2	Jones, M., Fret, J. and Smith B. (2010) 'Fatigue failures in carbon composites', *Journal of Materials Science*, 54, pp. 201–220	2. Year of publication in brackets. 3. Article title held within single quotes.
>2	Jones, M. D., Fret, J.B. and Smith B. (2010) 'Fatigue failures in carbon composites', *Journal of Materials Science*, 54, pp. 201. DOI: 12.1192/ 0003-1112/54/1/201 (Accessed: 28 March 2012)	4. Capitalise first letter of first word – rest in lower case unless referring to a proper noun. 5. Source of publication. 6. Volume or part number. 7. Page number(s). 8. Digital object identifier if available 9. Date accessed.
News Article	Burns, S. (2012) 'Biofuels to be used in local buses', *Glasgow Herald*, 20th March, p. 17	Note that the name of the source newspaper is italicised.
News Article	*Glasgow Herald* (2012) 'Infection crisis in local hospitals', 20th March, p. 17	
News Article	Low, D. (2012) 'Concrete cancer threat to bridges', 20th March, p. 17, *Herald UK* [Online]. Available at: http:// www.heraldnewspapers.co.uk/ (Accessed: 28th March 2012)	When referring to an e-journal or an article in an e-journal then the title of the source is italicised. The title of the article itself is presented in the standard font.

4.1.2   Books

The format for books is similar to that for journals but requires a little bit of thought. In many respects, a journal is in fact a book! Each article within it could be considered a separate chapter. Thus when you construct the reference to a journal article – you place the title of the article in quotation marks and present the name of the journal in italics. The same approach holds for books whereby the title is italicised (without quotes). Any chapter or section that has been specifically referred to is placed in quotation marks – just as you would when referring to an article within a

Table 4.3 Harvard formatting style for books.

Authors	Bibliographic style for books	Rules
1	Jones, M. (2012) *Carbon Materials*, 2nd edn. London: Egstrom Press.	1. Surname of author goes first followed by a comma and then the initials. Two authors should be separated by an 'and' with multiple authors by a series of commas and last two names by 'and'.
2	Jones, M. and Hardy, D. (2012) *Functional Materials*. New York: Blandbury Press.	
1	Jones, M. (ed.) (2010) *Fatigue failures in modern steel structures*, 3rd edn London: Engineering Open Press.	2. If the book has an editor then '(ed.)' is added after the name. 3. Year of publication is placed in brackets.
3	Jones, M., Smith, J. and Murray, K. (eds.) (2010) *Wound management in modern surgery*, 3rd edn London: Bioengineering Open Press.	4. Title is *italicised* but **no** single quotes. 5. Capitalise the first letter of the first word – the rest in lower case unless referring to a proper noun. 6. Edition (edn). (If first edition then ignore). 7. Geographical location of publisher followed by a colon. 8. Name of publisher. 9. Volume number (if appropriate).
1*	Buckfield, M. (2010) 'Electrolytic Modification of Carbon Steel', in Jones, M. (ed.) *Fatigue failures in modern steel structures*, 3rd edn London: Engineering Open Press. pp 201–212	When referring to an essay or chapter in a compilation then the format is a combination of journal and book style. The article title is held within single quotation marks, whereas the book title is not. The pages referred to should also be included.
0	*Oxford dictionary of chemical terms* (2010) Oxford: Oxford Independent Press.	Where there is no author then the same format as for books is used but simply omit the author.

journal. There are of course some minor additions such as the edition, where the book was printed and the name of the publisher. Examples of the format used when referring to books are highlighted in Table 4.3.

4.1.3 Electronic Publications

Electronic media is becoming increasingly common and will no doubt predominate in the near future. The format is very similar to that of the hard-copy material with the exception that you must specify '[online]' and provide the full URL detailing where the article can be found. More recent articles will have a digital object identifier (DOI) and this should be included where available. The latter has emerged in response to the fluidity of web pages within the internet where content can be moved from one URL to another as services are rationalised and updated – the DOI is a constant marker. The formats required for e-journals and web sources are detailed in Table 4.4 and Table 4.5.

Table 4.4 Harvard formatting conventions for electronic media.

Authors	Bibliographic style for electronic Journals	Rules
E-journals	Jones, M.D. and Hardy, D. (2012) 'Smart Bandage'. International Journal of Materials Science, 40, pp. 2–9 [Online]. Available at: http://www.IJMATS.com (Accessed: 28 March 2012)	This style of reference is typical of the many open-access publishers that are predominately internet based. This example is for an article from the web site of the journal.
E-journals	Jones, M.D. and Hardy, D. (2012) 'Smart Bandage'. Materials Science Journal, 40, pp. 2–9. Science Central [Online]. Available at: http://www.sciencecentral.co.uk/MSJ (Accessed: 28 March 2012)	Often, a publisher who has several publication titles under their management will place the articles within a central repository (i.e. Science Direct). The name of the repository should therefore be placed after the journal details.
E-journal Advance Articles	Smith, L. (2012) 'Hetero-atom inorganic cages'. International Journal of Materials Science, [Preprint]. Available at: http://www.IJMATS.com (Accessed: 28 March 2012). DOI 10.2033/0003-2324/40	Often a journal will promote forthcoming articles before their formal inclusion within the actual publication. There will be scant details other than the URL where the paper can be found. Occasionally, these "advance articles" will come with a DOI. When referencing such articles – state '[Preprint]' in the reference.
E books	Jones, M.D. and Hardy, D. (2012) Functional Materials. Blandbury Netforum [Online]. Available at: http://www.blandbury.com (Accessed: 28 March 2012)	Same initial format for standard hard-copy books. Additions to the format are: 1. Publisher is italicised; 2. [Online] added; 3. Available at: http:// . . . . . .; 4. (Accessed : . . . . . .).

Table 4.5 Harvard formatting styles for web-based sources.

Case	Example Bibliographic style	Rules
Web Page Report	Hawtree, M.D. (2012) *Functional Materials.* Available at: http://www.constructionmats.org/hawtree/ (Accessed: 28 March 2012)	Increasingly, reports from nonpeer-reviewed sources can be found online – especially from suppliers offering 'further details' on their products, charity information data-sheets, etc. Try to locate an author within the report – usually found
Web Page Download Document	Dunwoody, M. (2012) *Impurity Analysis* Available at: http://www.constructionmats.org/Impurity/AnalysisGuide.pdf (Accessed: 28 March 2012)	either on the title page or at the end of the report. If there is no date associated with the actual report then you should enter the date that the web page was last updated.
Organisa-tion Web Fact Sheets	Combat Diabetes (2012) *Diabetes Prevalence.* Available at: http://www.comdiab.org/CD-Stats2010/ (Accessed: 28 March 2012)	Often the report will be in the form of a downloadable file – typically pdf but can also be in document or spreadsheet format.
		If there are no clear authors then the name of the source organisation should be placed at the beginning of the reference. The format should be similar to that of books with the title of the report in italics with no quotes.
Web Page	http://www.myweb/constructionmats.html (Accessed: 28 March 2012)	Sometimes a search engine will direct you to a webpage that has been created by an enthusiast or interested party, but the amateur/ad hoc nature of its creation may preclude there being any name or identifiable date associated with it. In this case, the reference would simply be the URL and the date accessed. You should be very cautious about including these references and should, if possible, find alternatives.
Wikipedia	'Biofuel' (2012) *Wikipedia.* Available at: http://en.wikipedia.org/wiki/biofuels (Accessed: 28 March 2012)	There is no getting away from the wealth of information that is held within Wikipedia. It should, however, be viewed with caution as it is seldom peer reviewed and, as is the nature of wikis, can be modified and facts altered either by accident or through malicious endeavour – you should think carefully before citing a wiki and the weight that it gives to your argument in the main text.

4.1.4   Reference Order and Presentation Style

Now that you are accustomed to the individual formats of the references the question that must be addressed relates to the structure of actual bibliography. The references are simply placed in alphabetical order according to the first author. There are, however, some important pointers that you should consider when adding the bibliographical details. A common mistake is simply to continue in the same style as the rest of the text. Normally, you will be asked to present your literature review in a specific manner: font 12, double spaced, etc. The spacing allows your supervisor to read through and annotate your text 'between the lines'. It looks clumsy, however, when it comes to the bibliography.

Compare the standard lazy approach with the alternative in Table 4.6 where it is clear that you have put some effort into the presentation. The failings of the lazy approach are that there is no clear delineation between the references. It is hard to pick out the references at a glance. There is also little to distinguish the bibliography from the main text. The alternative approach makes use of a 'hanging paragraph' style whereby the first author is

Table 4.6 Different approaches to structuring a Harvard style bibliography.

Lazy approach	*Give me the marks approach!*
Bibliography	**Bibliography**
Bostrom, A. and Lofstedt, R.E., (2010) 'Nanotechnology risk communication past and prologue', Risk Analysis, 30, pp. 1645–1662	Bostrom, A. and Lofstedt, R.E., (2010) 'Nanotechnology risk communication past and prologue', Risk Analysis, 30, pp. 1645–1662
Cacciatore, M.A., Scheufele, D.A. and Corley, E.A. (2011) 'From enabling technology to applications: The evolution of risk perceptions about nanotechnology', Public Understanding of Science, 20, pp. 385–404	Cacciatore, M.A., Scheufele, D.A. and Corley, E.A. (2011) 'From enabling technology to applications: The evolution of risk perceptions about nanotechnology', Public Understanding of Science, 20, pp. 385–404
Dudo, A., Dunwoody, S. and Scheufele, D.A. (2011) 'The emergence of nano news: tracking thematic trends and changes in U.S. newspaper coverage of nanotechnology, Journalism & Mass Communication Quarterly, 88, pp. 55–75.	Dudo, A., Dunwoody, S. and Scheufele, D.A. (2011) 'The emergence of nano news: tracking thematic trends and changes in U.S. newspaper coverage of nanotechnology, Journalism & Mass Communication Quarterly, 88, pp. 55–75

clearly discernible – an important eye marker as the first author is invariably the memory aid the reader takes from the in-text citation when searching the bibliography. The format is reduced to a single space and each reference is effectively isolated and, as a result, is markedly easier to view. It is also an idea to reduce the font size to one or two points below that used in the main text – again to distinguish the bibliography as a separate, easily identifiable section.

4.2 THE NUMERIC SYSTEM

The numeric citation style tends to be more common within the physical and engineering sciences and has a simpler signposting style than the Harvard approach as the name–date system is replaced by a number. There can still be some resonances to the Harvard style where the work revolves around a single piece of work. Consider the examples of in-text citation presented in Table 4.7.

It would be a bit silly, not to mention clumsy, to use Example 1. A much better approach would be to use Example 3, but this should be done only if you are focusing on the work conducted by Dunwoody. You may be repeating it or attempting to build on the work that was initiated by the author. The third example requires a little bit more work in constructing the sentence, but is arguably, more appropriate to situations where you are making passing reference to the previous work.

4.2.1 Numeric Style Bibliographies

The specific format of the bibliography depends on which variant you have been instructed to adopt. Vancouver and the Modern Humanities Research Association (MHRA) are two of the more

Table 4.7 Different in-text citation styles.

Example	In-text citation
1	"Reference 1 demonstrated that a high anodic potential can induce the exfoliation of carbon fibre."
2	"Dunwoody (2010) demonstrated that a high anodic potential can induce the exfoliation of carbon fibre."
3	"It has been demonstrated that a high anodic potential can induce the exfoliation of carbon fibre (1)."

common. A quick examination of different journals, however, will reveal that the final format is very much dependent upon which journal you wish to publish. So what do you do? Again, refer to your supervisor. You might find that they don't really care what the final format of the bibliography is, but simply want the numerical citation style. In that case, the answer is to choose one and then, crucially, be consistent. There is nothing worse than seeing different citation styles littered throughout a dissertation and then capped off with a bibliography that changes style from one reference to another.

The nature of the citation remains the same except that where you make reference to other people's work you simply put a number – either as a superscript or encased with parentheses or square brackets. In the absence of any direction – choose the style that you think best suits your work. An important point to note, however, is the order in which the citations appear. The citation number must appear in logical sequence: 1, 2, 3, 4, etc. and not random such as: 1, 3, 2, 15, 6, etc. What do you do if you need to reference the work again – later in the dissertation? If you are citing a piece of work that you have already discussed then you simply use the same reference number. So let's suppose, when constructing your discussion, you have cited three works (1, 2, 3) and then need to cite another aspect of the work that is contained within reference one. The sequence then becomes: 1, 2, 3, 1. If you add another new citation, the number increments as normal and it would be introduced as the number 4. Consider the two examples presented in Table 4.8 of an introductory paragraph on crystallography.

The order shown on the left pane of Table 4.8 follows 1, 2, 3, 4, 5 with reference one being cited again between references 4 and 5. This is acceptable as reference one has been previously introduced. Now compare with the right hand pane in Table 4.8. The order of appearance is 1, 5, 3, 4, 2. Reference one has been cited twice but this is ok as it was introduced earlier – just as in the previous example. Reference 5, however, is clearly out of sync. The latter is a common mistake and tends to arise where students have found an important reference late in the day that must be added in to the text lest the examiner think they haven't studied the literature closely enough.

There are two options when you need to include a new reference within the text: substitution of an existing reference or insertion.

Table 4.8   Citation sequences.

Correct Citation Order	Incorrect/Random Citation Order
Introduction	**Introduction**
Materials science and engineering has evolved into a broad interdisciplinary combination of metallurgy and polymer science, inorganic and structural chemistry, mineralogy, glass and ceramic technology and solid-state physics (1) that includes advanced material applications such as: semiconductors (1,2), nanotechnology (3) and biomaterials (4). Materials science investigates the structure of materials at the atomic or molecular scale and relates this to the resultant macroscopic properties (1) and has increasingly become an additional component to most core engineering degrees in the UK (5).	Materials science and engineering has evolved into a broad interdisciplinary combination of metallurgy and polymer science, inorganic and structural chemistry, mineralogy, glass and ceramic technology and solid-state physics (1) that includes advanced material applications such as: semiconductors (1,5), nanotechnology (3) and biomaterials (4). Materials science investigates the structure of materials at the atomic or molecular scale and relates this to the resultant macroscopic properties (1) and has increasingly become an additional component to most core engineering degrees in the UK (2).

The former is by far the easier, whereas insertion will require the tedious renumbering of all the citations that appear after its introduction. The temptation (as shown in Table 4.8) is simply to insert it without any due regard to the numbering system. Again, this will shriek to the examiner that you are lazy and unprofessional. Re-numbering can, however, be very dangerous and requires considerable diligence as failure to increment the reference numbers can mean that a reader is directed to an incorrect reference if they follow a citation that has escaped your correction.

4.2.2   Vancouver Bibliography Format

There are a number of additional differences between the Harvard and numeric system – the main one being that rather than being listed in alphabetical order, the bibliography is structured with the references presented in order of appearance. The format is much simpler than the Harvard with little punctuation and furthermore, it is devoid of italics. The various styles for the different source materials are detailed in Table 4.9. The stylistic tips advocated in the Harvard system are equally pertinent here but rather than having the first author name offset – the reference number is offset instead.

Table 4.9 Vancouver style bibliographic formatting.

Source Type	Bibliographic style for electronic Journals	Rules
Books	1. Jones MD. Smart Bandage Technologies, London: Blank Publications; 2010.	The Vancouver style is much more minimalist than the Harvard in terms of formatting. A quick glance and you will see a marked reduction in the amount of punctuation. There is also no need to remember which part of the reference you should put in italics. There are, nevertheless, rules which should be adhered to:
Book Chapter	1. Donald K. Electrochemical oxidation of purines. In Paulson GH, Jolt KL, editors. Bioelectrochemistry of small metabolites. Glasgow: Central Press; 2010. p. 3–9.	
E-Books	1. Jones MD Wound Interventions [e-book]. New York: Wilson Publications; 2012. Available from WilsonScience. http://www.wilsonscience.com	1. Surname, then initials and separated only by a single space. No period after each initial. If there are more than six authors then only specify the first six and then terminate the list with *et al*. Remember to put the latter in italics.
Journals	1. Fleck D, Grant Y, Klasp L. Spectroscopic analysis of biofilms on metal substrates. Journal of Biological Spectroscopy 2011; 22:12–17.	2. The title is then presented but without any quotation marks.
Journals	1. Hart M, Fullton L, Crisp J, Mall H, Ventnor J, Hartness T *et al*. Drug release dynamics from a pH responsive polymer. Journal of Materials Technology 2012; 22:34–39.	3. For books: the location of the publisher – colon then the name of the publisher. This is followed by a semi colon and the year of publication.
E-journal Articles	1. Smith L. Inorganic cages for medical applications. International Journal of Biomedical Materials, [internet]. 2010; 32:23–29. Available from: http://www.biomedpubs.com/biomedmats/ (Accessed: 28 March 2012). DOI 10.2033/0003-2324/40	4. For journals: the name of the journal followed by the year of publication. A semi colon separates the year from the volume and a colon the page number. Note that in this style there is no pp. To denote pages – simply detail the numbers.
E books	1. Jones, M.D. and Hardy, D. (2012) Functional Materials. Blandbury Netforum [Online]. Available at: http://www.blandbury.com (Accessed: 28 March 2012)	

4.3 HARVARD *VS.* NUMERIC

You may well wonder why people use the numeric system. The advantage of this style over the Harvard relates to the fact that the latter tends to disrupt the flow of the text especially if there are a lot of sources, further compounded by sources with lots of authors. The two reference systems are compared in Table 4.10.

4.4 AUTOMATIC REFERENCING SOFTWARE

You may ask – do these exist? Yes! Why didn't we mention this at the start? For most essays you really don't need them – especially if you haven't used them before. Each has a learning curve and do you really have the time to wrestle the software into submission? The latter is a major consideration when you are at the critical point of compiling your dissertation. There are, however, a number of software options – in some cases your department/student learning resource centre may have a site license for them. The more common are Endnote (r), Refworks(r) and Reference Manager(r), a quick internet search will reveal myriad other options. The basic rationale is that they fit seamlessly within word processing packages such as Word and all you have to do is provide the reference system with appropriate literature references. With a few keystrokes or mouse clicks you can insert the citation marker within the text and the software will construct the bibliography according to the style you dictate. There is no doubting their usefulness – especially if using the numeric system as it will diligently re-number your text quietly, efficiently and without any loss of hair on your part should you be required to insert a new reference.

The downside, however, is that you have to develop the rhythm of putting all your references into the software. The WOK/WOS system described in Chapter 3 can export its search results in a format which is compatible with the software add-ins. For a normal project dissertation, it could be argued that it is just as easy to keep a list in the document and do the alphabetical ordering as you go. Is there a need for the software add-on? Probably not is the short answer. The software comes into its own in a professional setting where you are regularly preparing articles for publication as it will contain all the information on journal style and will speedily re-format the references accordingly at the press of a button.

Table 4.10   Harvard and Vancouver styles compared.

Harvard System	*Vancouver Style*
Introduction Nanotechnology has become an increasingly fashionable term in the media and is frequently portrayed within advertising promotions as representing a new panacea for improved performance – irrespective of the nature of the actual product or the "nano" modification (Ho, Scheufele and Corley, 2011; Dudo, Dunwoody and Scheufele, 2011). The significance of the term risks being trivialised but, more worryingly, the general ignorance of what "nano" really means can plant the seeds for future technophobia as witnessed by current concerns over the use of "nanosilver" (Cacciatore, Scheufele and Corley, 2011; Bostrum and Lofstedt, 2010; Siegrist, 2010).	**Introduction** Nanotechnology has become an increasingly fashionable term in the media and is frequently portrayed within advertising promotions as representing a new panacea for improved performance – irrespective of the nature of the actual product or the "nano" modification (1, 2). The significance of the term risks being trivialised but, more worryingly, the general ignorance of what "nano" really means can plant the seeds for future technophobia as witnessed by current concerns over the use of "nanosilver" (3–5).
Bibliography Bostrom, A. and Lofstedt, R.E., (2010) '*Nanotechnology risk communication past and prologue*', Risk Analysis, 30, pp. 1645–1662 Cacciatore, M.A., Scheufele, D.A. and Corley, E.A. (2011) 'From enabling technology to applications: The evolution of risk perceptions about nanotechnology', Public Understanding of Science, 20, pp. 385–404 Dudo, A., Dunwoody, S. and Scheufele, D.A. (2011) 'The emergence of nano news: tracking thematic trends and changes in U.S. newspaper coverage of nanotechnology, Journalism & Mass Communication Quarterly, 88, pp. 55–75 Ho, S.S., Scheufele, D.A. and Corley, E.A. (2011) 'Value predispositions, mass media, and attitudes toward nanotechnology: the interplay of public and experts', Science Communication, 33, pp. 167–200	**Bibliography** 1. Ho SS, Scheufele DA and Corley EA. Value predispositions, mass media, and attitudes toward nanotechnology: the interplay of public and experts, Science Communication 2011; 33: 167–200 2. Dudo A, Dunwoody S and Scheufele DA, The emergence of nano news: tracking thematic trends and changes in U.S. newspaper coverage of nanotechnology, Journalism & Mass Communication Quarterly 2011; 88: 55–75 3. Cacciatore MA, Scheufele DA and Corley EA From enabling technology to applications: The evolution of risk perceptions about nanotechnology, Public Understanding of Science 2011; 20: 385–404 4. Bostrom A and Lofstedt RE, *Nanotechnology risk communication past and prologue*, Risk Analysis 2010; 30: 1645–1662

Table 4.10 (*Continued*).

Harvard System	Vancouver Style
Siegrist, M., (2010) 'Predicting the future: review of public perception studies of nanotechnology', Human and Ecological Risk Assessment, 16, pp. 837–846	5. Siegrist M, Predicting the future: review of public perception studies of nanotechnology, Human and Ecological Risk Assessment 2010; 16: 837–846

There are merits to using it if you are completing a PhD where you could easily envisage your work being published in academic journals. If your department possesses a site license for the software, then it is certainly worth a look, otherwise we would recommend that for an undergraduate or Masters short project, manual referencing is probably just as easy and a lot cheaper.

4.5 SUMMARY/KEY POINTS

- ☑ Check your course guidelines for which format to use.
- ☑ Check with your supervisor that the recommended format is indeed what they prefer you to adopt.
- ☑ Remember to include the in-text citation where you make reference to other peoples work.
- ☑ Ensure that you have all the information required for the bibliographic reference.
- ☑ Remember to include 'Date accessed....' when including electronic sources.
- ☑ Ensure the format is consistent throughout your bibliography.
- ☑ Spend time to make the bibliography stand out.
- ☑ Double check the bulk text to ensure that everything that needs to be cited has been.

CHAPTER 5

Starting the Practical Work

5.0 GETTING STARTED

Oops...

Once the literature review has been completed, it is time to think about the actual project. Projects can vary tremendously and it is impossible to give a one size fits all framework as to how to proceed with the practical work. By outlining the objectives at an early stage, you can at least have an idea of the direction you are taking.

Research Project Success: The Essential Guide for Science and Engineering Students
Cliodhna McCormac, James Davis, Pagona Papakonstantinou and Neil I Ward
© Cliodhna McCormac, James Davis, Pagona Papakonstantinou and Neil I Ward 2012
Published by the Royal Society of Chemistry, www.rsc.org

Our previous recommendation, to skim-read the results sections of journal articles, should now begin to pay dividends as you should now have an idea of the type of experiments you need to conduct. Does this mean that you should simply disappear into the lab or glue yourself to your computer terminal and start coding? No. You need to speak to your supervisor! Since it's your project, shouldn't you be the one suggesting what to do? No – at least not so much in the early stages when the project is in its infancy. Returning to our Columbus scenario – he would have relied upon the local pilots to guide his ships out of the harbour and through the local waters until they hit the open seas. You have to do the same. You have to rely upon the guidance of your supervisor to get you going and heading in the correct direction. So, my advice at this stage? Patience!

5.1 HEALTH AND SAFETY

There will be a strong temptation to run off and throw yourself into doing stuff. This is not really the best idea since you will then be missing out on an important step. Health and safety regulations need to be addressed at the very outset, long before you do anything else. Some projects might not use chemicals or involve large dangerous bits of machinery – so does this exclude them from health and safety considerations? No. A risk assessment needs to be carried out, whatever the activity – whether you're dangling from a bridge inspecting corrosion of steel supports or sitting in front of a computer terminal writing an 'App'. In terms of definitions: a hazard is something that can cause harm, whereas a risk is the chance of it occurring. It is true that the more dangerous the activity the more complicated the form filling gets but, rather than viewing it as an arduous task – consider it as your safety net ensuring you avoid injury. Each department will have their own set style for such forms but, essentially, you need to consider the activities pertinent to the project, identify any potential hazards that are present or could arise and then consider how they could be minimised or removed. We have used the format that is currently used at the University of Ulster and attempted to strip the process down to the bare bones so that the principles can be adapted for the forms used at your institution. There are five basic components to completing a risk assessment irrespective of discipline or department:

1. Identify hazards associated with a given activity.
2. Decide who could be harmed.
3. Assess the risk and consider controls.
4. Document the assessment and date.
5. Periodically review data and update if required.

A list of the more common hazards liable to be encountered within the engineering and physical sciences are highlighted in Table 5.1.

Note that there can often be more than one hazard associated with an activity. Most documentation will account for this and all possible hazards should be recorded – not simply the ones you consider to be the most important. The next stage is to consider those that could be affected. Who would engage in the activity? At this point, you want to move away from the 'me' to a group description – so students, staff, visitors, etc. Now comes the tricky part. . .assessing the risk and suggesting control actions.

How severe could the injury be? What is the likelihood of it happening? These are the two questions that you need to address when attempting to gauge risk. If we take 'Severity' first, you need to think about the degree to which the injury will affect the person performing the activity. Three possibilities are usually considered and these are highlighted in Table 5.2.

Once the potential severity has been established, it is important to consider the likelihood of an accident arising (frequently, sometimes or rarely). It is then possible to make a judgement on the relative risk. One way of assessing it is to use the matrix outlined in Table 5.3.

An example of assessing an activity is shown in the abridged Health and Safety form indicated in Table 5.4. There are three tasks associated with the activity – essentially switching on the machine, filling it with a solvent and running a test sample through it. The hazards for each task are identified from those outlined in Table 5.1 and mention is made of the steps (controls) that can be adopted to minimise the risk. The risk rating is then assessed from the matrix in Table 5.3. The consequences of electrical shock are severe and could result in death, yet the risk rating highlighted in Table 5.4 is only considered to be 'Medium'. Why? The likelihood

Table 5.1   Common hazards.

General Hazards	*Specialist Occupational Hazards*
Burns	Asbestos
Dangerous machinery – entrapment, entanglement	Entry to confined spaces
Display Screen Equipment	Excessive noise
Drowning	Flammable/highly flammable material
Electric shock/burns	Ionising radiation
Excessive heat/cold	Lab/process waste
Explosions – air, pressure, steam, etc.	Lasers
Falls of objects from a height	Lead
Falls of people from a height	Nonionising, e.g. ultraviolet, infrared, microwave
Field trips – communication, sunburn, weather, cliffs, boats	Occupational stress
Fire	Radioactive waste
Fork lift trucks	Vibration
Gas/gas cylinders	Work on live electrical equipment
Glass	
Hazards associated with pregnancy/disability/young workers	
Inadequate lighting	
Inadequate space	
Inadequate ventilation	
Inherently dangerous tools, e.g. knives	
Lack of information/training	
Lack of supervision	
Lone working	
Manual handling	
Needles	**Chemical/Biological Hazards**
Ovens/furnaces	
Poor housekeeping	Asphyxiants
Poor maintenance	Cancer causing
Poorly maintained tools	Cells human/animal
Provision of unsuitable equipment	Clinical waste/hazardous waste
Refrigeration plant	Contact with animals
Restricted access and egress	Contact with poisonous plants
Sharp edges/scalpels/blades	Corrosive
Slips/trips/falls on the level	Flammable materials
Unsafe behaviour of individual	Harmful or irritant
Unsafe systems	Highly flammable materials
Unsuitable/inappropriate use of hand tools	Human blood/body fluids
Use of access equipment	Micro-organisms e.g. legionaella, hepatitis
Vehicles on premises and public roads	Sensitising
Violence/aggression	Substantial quantity of any dust
Welding–electric, oxyacetelyne	Toxic

Table 5.2   Classification of severity.

Severity	Description
Trivial	Minor injury that would require, at worst, general first aid
Significant	The injury would require specialist attention (burns, broken bones, etc.)
Severe	This is where the injury could to lead to permanent impairment or death

Table 5.3   Risk matrix.

		Trivial	Significant	Severe
Likelihood	**Frequently**	Medium	High	High
	Sometimes	Low	Medium	High
	Rarely	Low	Low	Medium
		Trivial	**Significant**	**Severe**
			Severity	

of it occurring is very low as most modern equipment will be engineered to prevent any access to the electronics contained within the grey boxes. Nevertheless, there is always the possibility that something could have gone wrong. Regular PAC testing is another control but this is only an annual test. Thus, the risk rating must be 'Medium'.

An added complication arises when using chemicals. This will inevitably require a separate form where you assess the potential hazards posed by the substance. Such forms are usually termed Control of Substances Hazardous to Health (COSHH) assessments and will document the nature of the activity and the chemical being used. There is a temptation to rely on the assumption that one

Table 5.4   Sample risk assessment form.

Activity Title	*HPLC Setup and Initialisation*		*Date*	*12/05/12*
Task	**Hazard**	**Controls required**	**In place?**	**Risk rating**
Configuration	Electrical shock	No exposed cables, regular PACT testing	Yes	Med
Sample injection	Needle stick	Autosampler system. Failsafe to prevent operation while sample door is open	Yes	Low
Filling of solvent reservoir	Spillage	Solvent contained in solvent tray with absorbent granules	Yes	Low

white powder is the same as the next. This is not the case! It is vital that you approach chemicals with a high degree of caution and don't simply rely on the word of a postgraduate researcher and don't accept glib statements of their assessment as being true. Do the legwork yourself. Whatever you do – make sure that you wear lab coat, safety glasses and gloves when using chemicals.

The label on the chemical will invariably contain a pictogram highlighting the threat that it poses – irritant, flammable, toxic, etc. Some of the more common are listed in Table 5.1, but this is only part of the story. You should try to determine what the exposure limits are – some chemicals are highly regulated – especially those that are cancer causing and those that can affect fertility and pose a risk to an unborn child. The question then is how do you find out what the hazards are? Chemicals stored in the lab should have a Materials Safety Data Sheet (MSDS) either in a folder or in electronic form. This is essentially the curriculum vitae of the substance you are about to use. It will document the form that it is in, its properties and the risks it poses. The only thing missing is its hobbies and interests. It is vital that you scan the MSDS of the chemicals you intend to use to ensure that you know exactly what the dangers are. The MSDS will contain sections that can include 'Hazard Statements', 'Precautionary Statements' and 'Risk and Safety' phrases. These statements are codes in which the nature of the hazard posed by the chemical are described. There is overlap between the various phrases that relates to changing legislative demands. The H (or R) phrase will inform you of the threat posed while the P statements will guide you in terms of prevention.

The controls employed to manage the risk will also depend on the quantity of material being used. The smaller the quantity the easier it is to manage the risk. It should also be possible to find the short-term exposure limit (STEL), permissible exposure limits (PEL), threshold limit value (TLV) and the time-weighted average (TWA) for an eight-hour shift. These tend to be more common for gaseous or particulate airborne substances and are usually expressed as parts per million (ppm) or milligrams per cubic metre (mg/m^3). These should, if known, be stated on the MSDS. The STEL value is the concentration that a worker can be continuously exposed to for short periods without suffering from irritation or tissue injury. It is usually expressed along with the TWA. The TWA is the average concentration to which you can be exposed over an 8-hour period. The STEL is often marginally higher than the TWA but remember – STEL is a short-term exposure and while there can be a number of occasions within an eight-hour period in which the STEL is reached – the total should not exceed the eight-hour TWA. The PEL is the maximum amount (or concentration) of substance to which a worker (or student!) can be exposed. The STEL, PEL and TWA are usually expressed as parts per million (ppm) or mg per cubic metre (mg/m^3). If you cannot locate a STEL, PEL or TWA value – do not assume the material is inherently safe. Where possible, try to keep chemicals contained – usually in a fume hood. If you can smell it, then you are exposing yourself to needless risk. In all cases, treat chemicals as if they are toxic – wear appropriate personal protection and make sure you clean up spills. The latter is almost inevitable – make sure you know how to deal with them and don't simply leave them for other people to deal with or ignore. Seek advice from your supervisor to find out protocol and do not assume you can wipe it with a tissue and drop it in the bin.

If you cannot find the MSDS within the lab – ask the technical staff and/or search for it on-line. The supplier of the chemical will normally have the MSDS catalogue on line. The Sigma-Aldrich® website contains an extensive section on the various definitions associated with H, R and P statements and you should consult this before starting to use any chemical. You must also consider how you will dispose of the chemicals once the study is complete. There will usually be waste chemical containers in labs that use chemicals on a regular basis. If not – ask the technical staff or your supervisor – don't simply throw the waste down the sink – especially solvents!

The form highlighted in Table 5.4 presents the bare bones of how to approach risk assessment and how to go about finding the information for the associated forms and procedures. Remember that they are all meant to help you so approach them with an open mind – irrespective of the task you are performing. You should keep a copy of the risk assessments and attach them to your dissertation in one of the appendices. Most employers are conscious of the need to be proactive in terms of Health and Safety and thus – rather than a chore – the initial form filling could be a good selling point in an interview situation.

5.2 NOTES

Make sure everything goes into your lab book. It is important that you watch out for note fatigue. Many students will start off in a blaze of enthusiasm and take notes on absolutely everything, but as the project progresses the diligence that was once prevalent tends to diminish. An attitude that questions the importance of noting every detail emerges and soon the lab books are bare or are filled retrospectively – usually the day before submission. The lab book is a treasure trove – it allows you to glimpse back in time and enables you to analyse where things went wrong and where they went right. Do not rely on your memory – it will mislead you.

5.3 ACQUIRING DATA

Most experiments within the engineering and physical sciences have the potential to generate screeds of data – whether simulation or practical 'hands on' experiments. The diversity of project in this field precludes us from providing a definitive guide as to what steps you should take in particular situations. Instead, what we hope to do is provide you with a basic framework for you to consult prior to embarking on the actual investigations.

5.3.1 Instrumentation

The majority of procedures within laboratories tend to be automated – irrespective of discipline, and it is inevitable that you will end up using some kind of equipment. Lab equipment tends to be subject to a type of conformity such that any piece of equipment deemed worthy of a place on a bench must be a grey or beige box with some buttons and a LCD display. Gone are the days when you would have had to use chart recorders and insert felt tip pens to record seemingly unintelligible squiggles. Data is invariably captured by an on-board computer and either presented directly to you via the display on the box or transferred to a computer. Before you dive in at the deep end and start pressing buttons with wild abandon, there are a few things you should consider:

- If you are unfamiliar with the machine – ask for guidance from a technician or postdoctoral researcher on how to use it properly. If possible, ask to be given the training while doing the literature review. This is not to say you start lab work, but it prepares you for stepping into the lab. It can also break up the routine of having to sit in front of a computer tracking down relevant articles all day.
- Find out if the instrument needs to be calibrated. Don't simply assume that it is ready to go and you can switch it on. Also, find out what to do if the system crashes – such things tend to happen when there is nobody about – or no-one who knows how to use it. Find out how to shut it down safely. You need to become an expert on how to use the machines. It is better to ask than simply trust the manual and it is better to be persistent in asking than be afraid to bother people.
- Find out what the limitations of the instrument are with respect to the work you are about to embark upon – essentially make sure that it can deliver what you are looking for accurately. Ask who else uses the instrument and seek them out to discover if they have discovered any quirks or limitations. This could save you a lot of time and heartache where you could have spent weeks trying, without success, to acquire a robust set of data.
- Find out how to get the data off the machine. Can you save the data to disk and if so can it be transferred to a spreadsheet program? Most instruments will have an option to export the data – either as an Excel file or as an ASCII file. Way back in the mists of time, you would have photocopied the output

from the instrument and simply stuck it into your report. If you have the options of getting a hardcopy – then do so – but rather than sticking it into your dissertation, stick it into your lab book.

- When you have the option of importing the data into a spreadsheet – you should be able to improve the graphical output – by this we mean change the scale, focus in on one particular area or combine multiple sets of data onto the one set of axes. Be careful, however, that when you are working with the data, you do not change the nature of it. It is one thing to import the data for the purpose of getting a higher-quality graphical representation, but it is another to selectively present data with the purpose of focusing on the 'good' stuff whilst deliberately cutting out the annoying other stuff that may be present and that you would rather not think about or discuss.

The classic example would be where you have been attempting to prepare and purify a new wonder drug. Imagine you pass it through a chromatography system to see what is present in your 'product' and find that there are a number of well-separated peaks. You could focus in on the main peak and paste it into your dissertation as proof that you had made the material. You must not, however, simply ignore the presence of the other peaks nor should you present only your 'refined' data. The latter is a falsehood and a lie by omission.

5.3.2 Lab Environment

It is important that you become familiar with how the lab works – in terms of procedures, policies and personalities. Every lab will have its own way of working and you need to observe how things are done. The best time to do this is when learning how to use the equipment. The training sessions will inevitably mean you are standing about – hopefully absorbing the relevant bits of information. During this period you are merely a visitor to the lab, a nonthreatening observer. The key phrase is 'observer'. Take the time, when you can, to watch how the lab functions in terms of the people working in it. On your first day in the lab – ask to be shown directly where things are, who uses what, what to do with waste, is there a rota for cleaning glassware, etc. Most importantly, ask where you are to be based and where you can store your materials. Before you start the actual work – remember – Health and Safety! Find out where the eye wash bottles are, first aid kits and who is the registered 'First Aider' in that lab. The last thing you want to do is to have an injury and not know who to seek assistance from.

5.3.3   Model Systems

Once you are confident that you know how the equipment works and what the procedures are within the lab environment, you can begin to think about starting your project. Where do you start? The best idea is to start with something that has been done already. You may ask what the point of that is when you are supposed to be pushing back the frontiers of science. It is great to be keen, but silly to think that you can plunge straight in and get stunning results. Actually, they may well be stunning – but not in the sense you had hoped. It is much more sensible to spend the first few sessions repeating previous work until you are confident in actually doing stuff! It is all very well being told what to do and what is supposed to happen, but these only form part of the learning process, whereas the process of actually doing it is where you gain the necessary experience. You need to think of the preliminary work as a model system whose results you should be able to reproduce. This should serve as a gauge through which to compare your effort, allowing you to familiarise yourself with the lab, the instruments and to boost your confidence.

Model systems are very much the foundation from which you should begin to build. In many cases you may be trying to develop a new technique or material. In order to assess the effectiveness of your discovery, it will be necessary to compare it against an established process or product. Thus, the model system not only accords you the opportunity to get to grips with the lab work, it provides some hard data that can be used later – with the outcomes from these initial experiments being included within your dissertation. They serve as an indicator to the examiner that you are competent and have done the necessary legwork but, at the same time, provide you with the all-important standard through which to present a critical discussion of the subsequent experiments.

5.4 DATA ANALYSIS AND PRESENTATION

Once you have started in the lab and are generating results, the next question is what to do with the data? You don't simply start doing 'stuff' with no game plan. There must be a reason for following a certain path and conducting an experiment. At the end of the experiment or investigation you must therefore decide if the data you obtain matches your expectations. This applies across the board, irrespective of project area. Consider the example below.

Imagine that the project you are embarking on requires you to investigate a new synthetic pathway for a drug that could be used to treat heart disease. The process involves five distinct steps in which the product from one step is used as the starting material for the next stage. It is extremely unlikely that you would complete each stage and then assume that it was a success and blindly continue on through the process. This contradicts the project management ethos outlined in Chapter 2. Remember that you must break a project into manageable parts and then assess the progress of the project once a stage has been completed. Therefore you should analyse each intermediate product to confirm that it is in fact the compound you were trying to make before continuing with the next step. In order to confirm the latter, you would perform some tests and then compare them against what you would expect to find.

In the above example, you could evaluate the spectral properties (typically infrared, nuclear magnetic resonance or mass spectroscopy) of the molecule that would give rise to structural information. The trick then is to note down in your lab book what the key properties are (peak positions, etc.) along with what you expected. This will prove invaluable when it comes to the writing-up phase. The worst thing you can do is to blithely assume that the data is correct without challenging its validity. All too often students present data without question and without any form of analysis or corroboration. They simply present a chart and say this is what they found, but give no explanation as to whether the data is what they expected or if not – why it wasn't. When you obtain the results from an experiment you must try to explain them with respect to the underpinning theory and relate their significance to what other people have found. This is where you must demonstrate to the examiner that you are capable of critical thought and not simply following a recipe.

5.4.1 Presenting Your Data

The presentation of your data in a format that can communicate the significance of the outcomes is an important factor in the construction of a good thesis. It is vital that you present your data clearly and correctly. There are two basic approaches to presenting data: charts and tables. You have probably prepared countless numbers of these during the laboratory sessions and assignments, but are you presenting them effectively? There can be an all too automatic reliance on spreadsheets, such as Excel, to generate a chart and then simply copy and paste it into a dissertation without due consideration as to whether or not it has been properly constructed. The following sections explore the art of presenting your data.

5.4.2 Charts

Charts can be used to display instrument responses or to summarise the data obtained from a series of discrete experiments. In either case, there are a number of design considerations that you should follow in order to maximise your haul of marks. The most common type of chart in the engineering and physical sciences is the x–y plot where the variable you are interested in is plotted on the y-axis. Two crucial mistakes made by students, especially when in a rush to complete a lab report or assignment, are where the axes are left unlabelled or where, although the titles have been specified, the units have been omitted. This should not occur in a dissertation and the secret is to spend some time preparing your graph once you have the data. There is a temptation to record data in your lab book and then forget about it until it comes to the write up – by which time you are under pressure to submit the document. You must get into the habit of polishing your data, if not at the time of acquiring it, then certainly shortly afterwards. There are a number of nuances to consider when selecting the various options for an x–y plot. If there are hundreds or thousands of data points (i.e. when extracting the ASCII data for a spectrum) then it is probably better to simply omit the symbol for each point and have a line connect the dots. The presence of a marker for each point becomes clumsy and can detract from or even obscure the data. Best practice guides for the construction of x–y charts are detailed in Figure 5.1.

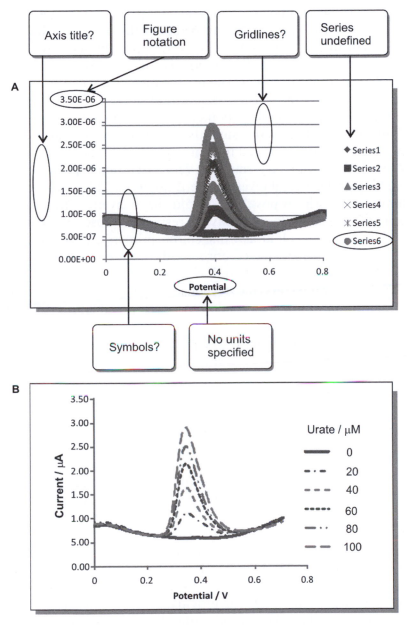

Figure 5.1 A) Common errors in chart construction, B) Recommended chart design.

There are several points to note in Figure 5.1. First, there is no chart title. This is really down to personal choice. You will rarely see a chart title in a research article so why present one here? Well, you could argue that a dissertation is different from an article. Should you decide to supply a chart title then please make sure you don't start it with the ghastly "A graph to show . . ..." It is obvious that it is a graph! The other points should be obvious. When specifying units please make sure that you adopt the 'divide by' approach rather than the all too common brackets style (i.e. '(V)'). When dealing with very small or very large numbers then most graphing packages will revert to the E notation. This tends to look clumsy and, where possible, should be converted to a more appropriate unit. When overlaying responses – use different line styles rather than colour as while colour printers are widespread – clarity can sometimes be compromised – especially if it's reproduced later on with a black and white copier.

Each line represents the response obtained for a particular concentration of urate. It would have been possible to display an individual chart for each but, it would have been much harder to compare. Thus, combining the responses into a single plot allows for easier comparison and it can be seen that the oxidation peak at $+0.35$ V increases with increasing concentration of urate.

When should you use symbols? The simple answer would be where there are a smaller number of data points. If we return to Figure 5.1B, then rather than only providing a direct comparison of the responses for each concentration, it would be good to show a more quantitative chart that highlights the variation in the height of the peak observed at $+0.35$ V with the concentration of urate. This would give rise to another x–y chart as indicated in Figure 5.2A. Note that the issues highlighted in Figure 5.1 have been taken on board and everything is correctly labelled. The data itself is a scatter plot in which only the symbols are shown. There is also a line on the graph, but it is the best fit line as determined by the spreadsheet's regression function. Linear relationships tend to be quite common and, where you anticipate such a relationship in the experiment you have been conducting, then you should make use of the 'trend line' function rather than simply connecting the dots. The latter will tell you nothing and looks amateurish. Ensure that you display the equation and the r squared value when using the trend line to indicate to the reader how good the fit is.

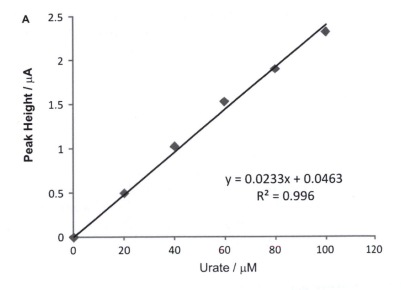

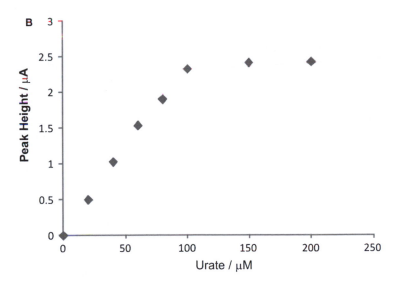

Figure 5.2 When and when not to use the trend line.

A word of caution – only use the trend line where you know or are confident that there is a linear relationship. In the example shown in Figure 5.2A, the concentration of urate is found to be linear over a given range of concentration. The danger is that you use the regression equation to calculate the magnitude of the peak height beyond 100 μM. When presenting charts with trend lines ensure that you document in the text the conditions under which linearity was observed. Do not extrapolate beyond the experiment data points as the behaviour could so easily change – in this case the response was saturated beyond 100 μM as indicated in Figure 5.2B. Thus, any calculation for the response at 200 μM, based on the regression equation, would have been massively wrong.

If there is no clear pattern between the x and y data, then present only the data points and resist the temptation to select the symbol plus line option that serves only to join the dots. It is worth noting that most spreadsheets will have the option of nonlinear regression – but be wary of how these are used and adopt them only when confident that you understand how they work.

5.4.3   A Word about Error

Throughout your investigation you must remain constantly vigilant for sources of error – these can fall into two broad categories: systematic and random. The latter tend to generate a constant bias in your results – usually as an instrumental or procedural artefact and typically where the instrument has not been calibrated properly. Random errors can be attributed to numerous sources, but most are a consequence of being human. Lack of concentration is usually the demon to watch out for here. You must try to minimise errors through regularly checking your equipment and being more careful and conscientious in the lab. Errors cannot be eliminated, but they can be minimised through good laboratory practice.

When you have completed one experiment, it is unlikely that you would simply stop there and consider that to be the best result you could hope for. It is common for an experiment to be conducted at least in duplicate and preferably in triplicate. The repetition of an experiment allows you to quantitatively estimate the error within the process you are using or developing. Most spreadsheet packages will allow you to quantify the degree of error in your investigations. Averages, range, standard deviation are the usual approaches. Error bars are another option. These tend to be more prevalent in Life Sciences dissertations than in the engineering and physical sciences but, if possible, you should consider their inclusion as it will add considerable weight to the professionalism of the final report. This is assuming, of course, that you take the necessary steps to obtain accurate and reproducible results.

Error bars can take many forms and different types can provide different types of information. There are four general approaches to constructing error bars:

- Range;
- standard deviation;
- standard error;
- confidence interval (usually 95%).

The error bars, employing range or standard deviation, are generally termed descriptive as they give you an indication as to how the data is spread (Box 5.1). The remaining two are generally classed as inferential and are typically used when comparing samples from different groups. You might be comparing the effectiveness of an antimicrobial drug you have designed with a nonactive (placebo) variant. Essentially, you would use the latter two when comparing experimental results with controls. The choice is dependent upon your particular project, but be careful not to simply pick one at random. Once selected, you must specify which type of error bar is being used and, crucially, the number of times the experiment has been repeated.

BOX 5.1

When using error bars you must ensure that you clearly specify which type of error bar you are presenting (range, standard deviation, etc.) and the number of repeat experiments. Remember that a repeat experiment means just that – the entire process is completed and not simply the same sample run three times. The latter will only provide you with an estimate of the error involved in the actual measurement – not the experiment.

This book cannot give you a tutorial on the use of statistics – the field is too broad and there already exists a wealth of excellent books dedicated to the use of statistics in various disciplines. Rather, our intention is simply to draw your attention to some of the misconceptions that routinely arise in undergraduate dissertations such that you do not fall foul of them.

5.4.4 Equations

It is almost inevitable that you will perform some form of data manipulation – usually through plugging the numbers you get from the various machines into some form of equation in the hope of getting more meaningful data out. Do you need to define the equations that you use? This question harks back to the section on referencing where there was a dilemma as to whether or not common knowledge required a citation. A similar answer arises here in that if the equation is well known within the community in which the project is based, then probably not. Consider for example the case where you are reading for a chemistry degree – would you need to define the equation that is used day in day out by chemists to calculate the number of moles of a substance in a given solution with a known concentration. No. If, however, you were studying engineering then it may be necessary. The more specialist or complex the equation becomes then the more appropriate it will be to present it and define the corresponding terms. What are the issues that surround presenting equations? They must be numbered and introduced in the text as you would a figure or a table. You must also define the terms (abbreviations) used and the units, as highlighted for the Nernst equation in Figure 5.3.

Where there are only a few terms then these can often be accommodated in the main text rather than listing them as indicated in Figure 5.3.

5.4.5 Tables

The formatting characteristics of tables have already been considered in Chapter 3. In this context, however, they are usually employed to summarise the final outcomes for a series of discrete experiments or, for example, to compare the properties of one substance with another (i.e. melting points, peak positions, etc). Use tables to summarise and support your discussion.

Let's imagine that you spent weeks in the lab preparing a wonder molecule. You would then perform the tests required to

$$aOx + n\bar{e} \leftrightarrow bRed \qquad \dots\dots\dots\text{Eq. 5.1}$$

$$E = E^{\circ} - \frac{RT}{nF} \ln \frac{a_{red}^{b}}{a_{ox}^{a}} \qquad \dots\dots\dots\text{Eq. 5.2}$$

where:

E = *Electrode potential (V)*
E° = *Standard electrode potential (V)*
R = *Gas constant (8.314 J $K^{-1}mol^{-1}$)*
T = *Temperature in Kelvin (K)*
n = *Number of electrons*
F = *Faraday constant (9.649x10^{4} C mol^{-1})*

a_{ox} *and* a_{red} *are activities of oxidized and reduced species.*

Figure 5.3   How to present an equation within the text.

confirm that you had indeed got the correct molecular structure. There would be an obvious temptation to simply show the spectra and state in the text something along the lines of "the nmr spectrum is highlighted in Figure 4.3". This is really very, very bad. It is incredibly vague. There is no critical discussion to explain why the spectrum confirms that you have actually produced the desired molecule. The upshot is that the examiner has to study the spectrum themselves and try to decide whether or not what you have said is true. It would be much better if you added in a table that summarises the various groups expected (predicted) and the peak/ positions that have led you to the conclusion that you have actually made it. The examiner can then glance at the spectrum and the table and quickly make an assessment without having to spend the time re-analysing your data.

Remember that it is important that you make your dissertation a pleasure to read. You do not want to litter the pages with poorly presented data and, worse, data that has not been analysed properly nor critically appraised. Go through the section and ask yourself the important questions – "Have I done enough to explain the results?" If you are not sure what a given set of results is telling you – ask your supervisor to help you interpret the data. It may be that it is rubbish – equally, it could be quite brilliant. How will you know if you simply say in your dissertation "this is what I observed"? You must demonstrate critical thinking.

5.5 SUMMARY/KEY POINTS

☑ Ask your supervisor to talk you through the first experiments so you are aware of what needs to be done and what the outcome should be. If unsure of any aspect – ask!

☑ Get into the habit of assessing the risks – keep a track of risk forms.

☑ Ask for guidance on how to use the equipment – become an expert.

☑ Use previous studies as model systems to help familiarise yourself with the techniques, lab and people.

☑ Where possible – obtain electronic copies of the data.

☑ Polish the charts and tables after acquiring the raw data and don't leave everything until the end of the project.

☑ Analyse your results as you get them and critically compare them with theory and/or previous studies. Remember to reference your work.

☑ Don't rely on performing a test once – repeat the experiment to obtain better results and to assess error.

☑ Take copious notes and write up your experiments as you go.

CHAPTER 6

Constructing the Dissertation

6.0   THE DISSERTATION

Bet you wish you had taken proper notes!

This is the main outcome of the project or the end product. No matter what the scientific aim of the project is, the production of an excellent dissertation is the overriding aim as this is where the majority of the marks lie. All the work that is done in the library, the lab or at the computer goes into the pages of the dissertation,

Research Project Success: The Essential Guide for Science and Engineering Students
Cliodhna McCormac, James Davis, Pagona Papakonstantinou and Neil I Ward
© Cliodhna McCormac, James Davis, Pagona Papakonstantinou and Neil I Ward 2012
Published by the Royal Society of Chemistry, www.rsc.org

but they will mean little if the actual book is poorly constructed. You should start the dissertation at the beginning of the project and build it up as you progress so that, when the deadline for submission approaches, all it requires is a bit of polish. We have already outlined the virtues of doing this in Chapter 2. The aim of this chapter is to guide you through the various components that must go into it and to give some tips on how to make it sing out to the examiner "GIVE me the marks!"

It is likely that you will have been given instructions as to how the dissertation should be constructed in terms of layout/format, etc. (Box 6.1). Irrespective of subject, dissertations tend to have a common structure: there will be four core chapters although sometimes the Results and Discussion section are split into two separate chapters.

BOX 6.1

The conventional structure is as follows:

- Declaration – Certificate of Originality
- Abstract/Summary
- Acknowledgements
- Contents
- Chapter 1 – Introduction and Literature Review
- Chapter 2 – Methodology
- Chapter 3 – Experimental Details
- Chapter 4 – Results & Discussion
- Conclusions/Future Work
- References
- Appendices

The headings are pretty much self-explanatory, but a brief description of what goes in each is given below:

6.1   DECLARATION OF ORIGINALITY

This is usually a page that states that the work detailed herein is your own and that it hasn't been submitted elsewhere. Each university will have its own template and will usually require that you to sign the declaration. This is to ensure that you accept responsibility for the material presented within the dissertation and are aware of the consequences should you try to pass off other peoples work as your own (plagiarism is considered in Chapter 3).

6.2 THE ABSTRACT

This is a brief summary of what you set out to do (the aims) and what you have done. Note the tense. This must be in the past tense and take the third person. The entire dissertation should be written in the third person therefore no "I found that...", instead you must structure your sentences along the lines of "It was found...", etc. You should refer to the objectives (the milestones set out in Section 2.2) and proceed to document the outcomes of each. You are reporting the outcomes – as fact – along with a brief critical appraisal of their significance. The latter is essentially where you give the reader an indication as to whether it was successful or not. You may have been told that the abstract is to be a certain number of words. If not, then it should certainly be no more than one page. It must be concise. No experimental details unless they are critical to the outcomes – simply report the results observed and the outcome of your critical appraisal. Consider the example abstracts shown in Figure 6.1 and Figure 6.2. Remember the tale of Goldilocks – aim for an abstract that is neither too big nor too small. It should be 'just right'.

The first abstract is too short on detail, whereas the second contains a more expansive account of what the aim of the project was, the results obtained and their overall significance. There is some experimental detail in the abstract, but this is not trivial bulking out but rather the optimised conditions. These are the outcomes of the project and not simply the default settings on the instrument. The latter should not be appear in an abstract. An important point to note is that there should never be any references within an abstract.

6.3 ACKNOWLEDGEMENTS

This is a personal statement where you thank those that have helped you during the project – typically your supervisor, technicians, collaborators, etc. It can be tempting to thank everyone who

Abstract

The exploitation of 2-bromo-1,4-naphthoquinone (NQBr) as a selective redox label for the determination of reduced thiol functionalities (RSH) has been investigated. The system is selective for RSH functionality, giving distinct voltammetric signals for glutathione and cysteine but can also be adapted for broad spectrum thiol detection. Ion chromatographic protocols based on conductimetric detection enable total RSH analysis (protein and monomolecular moieties) within human plasma. Bromide released through the reaction can be easily quantified and integrated within normal IC measurements. The efficacy of the approach has been assessed and the response validated through comparison with the standard colorimetric technique.

Figure 6.1   Simple abstract devoid of any real detail.

has supported you thus far. It is wise, however, to avoid putting anything that will make people cringe. As noted in the beginning of the chapter – the dissertation has a life of its own and it might not just be you or your family who will see it but potential employers or students who take up the work in subsequent years. It is always a good idea to acknowledge your supervisor no matter how the project has gone on a personal level. The personal statement is not the place to maintain a grudge.

6.4   LIST OF ABBREVIATIONS

An alphabetical list of abbreviations used throughout the dissertation should be included at the beginning to which the reader can refer if necessary. Remember that the reader may not be an expert in the field – especially if it is a student following on from your project. The abbreviations must be defined within the text on the first occasion they are presented. For example:

"The limit of detection (LOD) is a major factor in the selection process and it is vital that the instrument is capable

Abstract

Plasma thiol (PSH) concentration has long been recognised as a potential indicator for assessing the severity of oxidative stress processes within physiological systems. While such measurements are normally restricted to research studies, this project sought to develop and characterise a novel approach through which this parameter could be exploited within routine clinical settings. The protocol was based on the rapid derivatisation of reduced thiol functionalities (protein and monomolecular moieties) through the homogenous reaction of a naphthoquinone bromide derivative.

Bromide released in the reaction was easily quantified through ion chromatography (Isocratic Dionex DX-120) utilising an IonPac® AS14 anion exchange column and a 25 mL sample loop. The released bromide was measured with conductivity detector. The mobile phase was optimised and consisted of sodium carbonate/bicarbonate (3.5 mM/1 mM) at a flow rate of 1.5 mL/min). Method selectivity and sensitivity was critically evaluated and the technique found to cover a clinically relevant range (15 μM–3.5mM PSH) with a detection limit of 9 μM PSH.

The efficacy of the approach for the analysis of human plasma from five volunteers was assessed -ranging from 49 to 72 μM with an intra assay variation of less than 5% in all cases. The responses were validated through comparison with the standard Ellman colorimetric technique. The analytical accuracy coupled to a run time of 5 min provides a fast method of measuring PSH concentrations that could be adopted within conventional clinical biochemistry settings and would provide a more convenient alternative to sending samples to specialist laboratories.

Figure 6.2 More comprehensive abstract.

of reaching below the minimum concentration of nitrate expected to be found in natural water. In addition to the system's LOD, cost is the other driver . . . . . ."

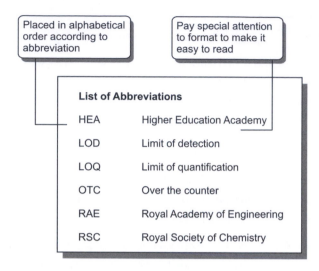

Figure 6.3   Abbreviations – style guide.

When introducing abbreviations – make sure they follow the word. Also try to avoid starting sentences with an abbreviation. The typical format of the list is shown in Figure 6.3.

6.5   LIST OF FIGURES

In addition to the abbreviations, there should be a page that lists the figures used throughout the work, along with the associated legend and the page upon which it could be found. Again, this could appear tiresome, but you are trying to make every effort to ensure that the dissertation is easy to read and to demonstrate to the examiner that you have gone the extra mile to make it a professional document.

6.6   CONTENTS AND PAGE NUMBERING

This can appear to be self-explanatory, but there are some conventions that should be considered. Chapter and section headings should be married with the corresponding page numbers – this much is obvious and hardly needs explaining. Chapters should be designated as 1, 2, 3, etc., with subheadings as 1.1, 1.2, 1.3, etc., and subsections as 1.1.1, 1.1.2, 1.1.3, etc. The difficulty arises when

Contents

Declaration of Originality..........	i
Abstract....................................	ii
Acknowledgements..................	iii
List of Abbreviations.................	iv
List of Figures..........................	v
Chapter 1	
1.1 Introduction & Aims............	1
1.2 Background........................	2
1.3 Existing Technologies........	6

vi

Figure 6.4 Typical format of the contents page.

considering which page to designate as page 1. One answer would be to say that surely it is the very first page of the report. No! Page one should be the start of the actual report and not the ancillary pages that precede it. Thus, the start of the introduction in Chapter 1 becomes page 1. A sample outlay for the contents section is outlined in Figure 6.4.

The order in which the ancillary pages appear is very much down to personal choice and the sketch shown in Figure 6.4 should only be considered as one option. The page number, however, is the important point in that, irrespective of order, Roman numerals should be used – and should also include the pages used for the contents pages themselves. Chapter 1 should be page 1. The front cover has no page number attributed to it. Corroboration for the type of format used can be obtained simply by looking at your textbooks.

6.7 CHAPTER 1: INTRODUCTION AND LITERATURE REVIEW

This chapter is where you begin to document the journey. Its construction has already been discussed in Chapters 3 and 4.

6.8   CHAPTER 2: METHODOLOGY

This section explains the approach you are going to take and will explain the theory behind what you are trying to achieve (Box 6.2).

> **BOX 6.2**
>
> This is where you demonstrate to the examiner that you know what you are doing. It should also be helpful to you as having to write out the theory should reinforce in your mind the science that underpins the work you have done.

Thus, chapter one presents the work that has been done before and highlights the weaknesses. This chapter sets out to explain how you will overcome those and justify why you have chosen a particular experimental design. This section should not contain any specifics with regard to the experiments that were carried out. It should contain the theory and appropriate descriptions of the techniques that were be used, etc. and, like Chapter 1, it should be fully referenced. The chapter should ideally end with a section that reiterates what the aim is and details the objectives (see Chapter 2).

6.9   CHAPTER 3: EXPERIMENTAL DETAILS

This is where you provide a summary of the materials used, the various instruments and their settings. It should also contain information on any specialist procedures undertaken in the course of the work and the level of detail should be sufficient to allow another student to repeat the work that you have done. Example calculations used in the preparation of materials could be inserted here or moved to an appendix. Note that it should not contain a mass of repetitive calculations that you have amassed over the course of the work. These will be detailed in your lab notebook. Rather, a formula or single example should be presented. When specifying materials and instruments, ensure that you record the manufacturer and their location along with the grade or instrumental model no. These are normally contained within brackets after the material or instrument. If a large number of materials or

Materials

Reagents used in the project are specified in Table 3.1. All were used as received without any further purification. Unless specified otherwise, all solutions were prepared on a daily basis using deionised water from an Elgastat (Elga, UK) water system.

Chemical Name	Quality/Grade	Supplier
Acetic Acid	Lab Reagent	Fisher Chemicals
Phosphoric Acid	Lab Reagent	Fisher Chemicals
Boric Acid	Analytical Reagent	Aldrich
Potassium Chloride	Analytical Reagent	Aldrich
Sodium Hydroxide	Analytical Reagent	Aldrich
Uric Acid	99%	Lancaster

Table 3.1 Chemicals used during the project

Instrumentation

Electrochemical measurements were conducted using a μAutolab Type III computer controlled potentiostat (Eco-Chemie, Utrecht, The Netherlands) using a three electrode configuration. Glassy carbon served as the working electrode (3 mm diameter, BAS Technicol, UK) with Platinum wire as the counter electrode. A 3 M NaCl Ag | AgCl half cell reference electrode (BAS Technicol, UK) completed the cell assembly. Electrochemical Quartz Crystal Microbalance (EQCM) measurements were obtained using a computer controlled Quartz Crystal Microbalance (Maxtek INC, USA) and polished 5 MHz Titanium/Gold crystals (Maxtek INC, USA).

Methods

Preparation of Buffer Solutions

Buffer solutions were prepared using a base mixture of acetic, boric and phosphoric acids (each at a concentration of 0.04 M) which was adjusted to the desired pH through the drop wise addition of sodium hydroxide (2 M).

Figure 6.5 Experimental details.

chemicals have been used then it may be an idea to provide a table of compound name, purity and supplier, etc.

6.10 CHAPTER 4: RESULTS AND DISCUSSION

This chapter can, depending on supervisor/departmental policy, be split into two separate sections whereby you simply report the results first and then discuss their significance in the subsequent chapter. The split approach can be useful if there are one or two experiments that generate lots of data. This can be presented and number crunched first and then discussed in the next/subsequent chapter. In the engineering and physical sciences, more often than not, there can be a series of experiments each leading on from the one before – almost in a linear progression. In this situation, it is easier to present the data for each set of experiments, critically analyse their significance and from there – lead onto the next part. In essence, it's like following a map, step by step. Returning to our Columbus theme – he didn't set out to discover a new trading route to the east all in one big voyage, rather there were lots of intermediate steps that led on to a discovery far greater than his original plan. This approach mirrors the fate of many projects – though the scale of the discovery may be more modest in many cases.

The chapter should therefore be broken down into subheadings that outline the various stages and will usually mirror the objectives outlined in Chapter 2. In each case there should be a brief paragraph to introduce what you have done and why before presenting the results. The presentation of the results has been discussed in Chapter 5. The subsection is not complete until you add in some critical appraisal of those results (Box 6.3).

BOX 6.3

Emphasis should be on discussion and interpretation of your results. This section should be clear, concise and critical. This is a key section since it provides a means for you to demonstrate your understanding of the subject and ability to cope with research investigations (with the minimum of supervision).

The common mistake is to simply state what you have found and leave it there. The phrase "I did this and I got that" is rather sadly exemplified in too many dissertations. It is important for you to place the results in context and this is where your skim reading of the articles woven into your literature review comes to the fore. You should have an appreciation of how your work relates to that of other groups and you should demonstrate that knowledge to the examiner. Are your results better or worse than expected? Are they better or worse than the work published by other groups?

If worse – why? If better – why? Do the results fall in with theory? If not – why? Look back on what you have written and see if the 'why?' word comes to mind. If it does then you need to put in a sentence or two to explain what you found.

6.11   CONCLUSIONS

This section can cause a bit of a stir as most students feel that they don't know what to write – after all they have said it all during the discussion sections in the previous chapter. The sense of trepidation is increased when they wonder how they will manage to fill this chapter. The previous chapters will have run to many pages yet they feel, mistakenly, that the conclusions should follow that trend. In reality, the conclusions section of the dissertation should only be a few pages. Even so – what goes into it?

The first steps should be to briefly remind the reader of what the problem was and its significance. A significant amount of time will have passed between starting to read the dissertation and reaching this point. It is always good to remind them of the main aim. The introductory portion in the first page should not, however, be simply cut and pasted but rather condensed in to a few sentences that highlight the issues.

Next – remind yourself of the objectives of the project. Use these as the framework through which to construct the remainder of this section. Do not present them as separate headings but merge them into the text. There should be no experimental detail – it is only a matter of cementing in the mind of the examiner what has been achieved in relation to those objectives that you set out to meet. What happens if some of the experiments failed? These results are no less valuable than the successful experiments, but only if you can give an account of the reasons why they were deemed to be a failure. It is no use simply stating that they failed and then abruptly going on to the next one. If possible, don't present an experiment as being a failure, but spin it round to find something positive – sometimes it is just a matter of perspective. Once the objectives have been discussed you should point out what the real-world significance of the outcomes is.

The final part of the conclusions section is where you speculate on how the investigation can be extended and what the ultimate outcomes could be. You are free to suggest alternative routes of enquiry to achieve the aims presented above. It is also possible to highlight here discoveries that, although not relevant to the present project and hence not pursued here, could be transferred to other fields.

6.12 BIBLIOGRAPHY

The structure of the bibliography has been considered in detail in Chapter 4. Remember – be consistent in your formatting. Make sure there are no missing references.

6.13 APPENDICES

This is the equivalent of the storage container that is only opened when looking for something specific. It is usually unmarked but is useful for containing raw data. The results section should highlight the processed data – either in the form of abridged tables or figures. This does not mean that you can throw out the raw data. There could be a mistake in your data analysis or a subsequent investigator may want to make use of it. The appendices can also be used to include supplementary material such as risk assessments and poster presentations etc.

6.14 SUMMARY/KEY POINTS

- ☑ Compile list of abbreviations.
- ☑ Compile list of figures.
- ☑ Insert page numbers starting from page containing '1.0 Introduction'.
- ☑ Use Roman numerals for all preceding pages.
- ☑ Remember to write in the third person – no I, me, we or our.
- ☑ The abstract should be informative and highlight the main achievements.
- ☑ Avoid being too sentimental in the acknowledgements.
- ☑ Spend time on the format of the contents page – make it easy to read.
- ☑ Be consistent in the format of the main text:

    Decide on font style for main text, heading, etc.
    Section heading should be in a larger font than subheadings
    Use bold to highlight – never use underline

☑ Remember to explain your results and refer back to the literature review to place your results in context.

☑ Summarise the main outcomes in the conclusions section and dare to dream as to how the work could be expanded or transferred to another research area.

☑ If using the numerical referencing system – make sure that number sequence is correct.

☑ Place miscellaneous data in the appendices.

CHAPTER 7

Posters and Oral Presentations

7.0 PRESENTING YOUR WORK

"That concludes my presentation.
Does anyone have any questions?"

Most courses will require you to present your project data in the
form of a poster or to give an oral presentation summarising

Research Project Success: The Essential Guide for Science and Engineering Students
Cliodhna McCormac, James Davis, Pagona Papakonstantinou and Neil I Ward
© Cliodhna McCormac, James Davis, Pagona Papakonstantinou and Neil I Ward 2012
Published by the Royal Society of Chemistry, www.rsc.org

the work you did. Some courses can appear to be of a particularly sadistic bent and will ask you to do both! Is this a bad thing? Your mind is probably screaming – YES! Pause for a minute......and breathe.....then think.....is it really so bad? Isn't it actually in your best interest? These tasks are set to help you gain confidence. You may well ask how standing up in front of everyone for fifteen nerve-wracking minutes is going to help your confidence! The answer is Practice. The confidence comes with experience and the more you do something, in principle, the better you should become. You may also think that what is the use of presentations when you only want to get a job and earn some money – you don't want to teach therefore why all this emphasis on presentation skills? The answer to that is simple: most jobs whether at the interview stage or in the work place demand that the employee possess a high level of communication skills. It is a question that is present on every reference a lecturer has to fill out for their students and hence the need to prepare you for the world outside the university. As we have mentioned throughout this book, the project does have a life beyond the university and will follow you, at least in the early stages of your career. Interviewers will ask to see the dissertation and presentations on the day are not uncommon.

This section cannot do much in terms of removing the sense of fear, dread and terror that can arise before giving a presentation, but it should give you the procedural tools required to quickly build an effective presentation – even at a moment's notice. The procedures outlined here are transferable and, hopefully, through knowing that the content of the poster or talk is fine, the fear of actually delivering the material can be reduced.

The crucial concept to keep in mind when constructing a presentation – poster or oral – is that you need to tell a story. The key to telling a good story is to engage the audience. You need to capture their interest so that they want to know what comes next. Think of the classic fairytale scenario where the princess is captured by the dragon, the knight sets off to kill said dragon and rescue her. The story culminates in the....knight being toasted. No! That would be a pretty rubbish story. Audiences are almost pre-programmed to want a happy ending. The princess has to be rescued! Thus, when it comes to preparing your presentation you need to:

- Set the scene to engage your audience – outline what the problem is and emphasise why it is important to the audience.
- Highlight the challenges faced and how you planned to overcome them.
- Describe what you actually did and found.
- Present the outcomes in a positive light that answer some of the problems highlighted in the introduction.

You already have all the information you need to prepare the presentation as you should have completed your dissertation. The challenge, however, is how you transfer a 50-page book onto either a single slide (poster) or a series of slides (oral). This is not such an arduous task as it first appears – after all, you have already created the framework for either type of presentation.

7.1 CRITICAL COMPONENTS

In essence, the presentation is very much like an abstract of the work you have done. If you paid attention to the instructions in Chapter 6 relating to the construction of the abstract, then you will be off to a flying start. You need to have distinct sections for: Introduction/Background, Aims, Methodology, Results and Conclusions. In some cases you may wish to add experimental details, but this should only really be included if the project involves tweaking the setup to optimise something. A similar case was considered in Chapter 6. The difference between a poster/oral presentation and the aforementioned abstract is that the former will contain selected graphical representations (schemes, pictures, charts, tables, etc.) to corroborate the claims you made. Keep things simple. This may seem obvious and I am sure you get the point, but it is worthwhile keeping that phrase at the forefront of your mind. You only want to present the main points.

Let's consider the various sections in more detail:

- Introduction/Background – this should be a modest para-graph of summary points that convey the significance of the project you are embarking on. You need to try to engage the curiosity of the audience much in the same way as did the first page in your Introduction. It is often referred to as the 'hook' through which to capture the audience. Use hard data to lend credibility but remember to reference prior work.

- Project Aims – these should state what it is you are going to do to solve the problem highlighted in the introduction.
- Methodology – a brief explanation of the approach you are taking – this again could be a paragraph, summary points or a schematic. Remember – try to minimise the use of words – use a picture/scheme where possible.
- Results and Discussion – this section should form the bulk of the presentation and should present the data obtained either in table or graphical format. Make sure all aspects of charts are legible from a distance. Do not try to put all your data on the poster by reducing the size of the graphs and squeezing them in. Where you have a lot of charts or spectra – use only the ones that highlight the main findings.
- Conclusions – Simply highlight the main outcomes of the project. This can be a simple paragraph or bullet points but must convey to the viewer what the significance of the results obtained are in the real world. Did you succeed? Make sure there is a positive spin on the summary.
- Bibliography – Use a smaller font – size 12 – to provide references as appropriate.

The mark scheme for a presentation will normally be divided into: Aims, Background, Scientific Content and Grammar/Format/Style. There will also be some marks assigned to your ability to answer questions on the material you have chosen to present. You may find that despite going to amazing lengths to create a truly impressive presentation in terms of its design, you still end up with a poor mark. Why? The bulk of the marks will normally be allocated to the 'Scientific Content'. Thus, particular attention must be paid to the Methodology and Results sections. Too often, students will bias the content towards the introduction. Remember, even if some of the experiments have not worked out as planned – try to highlight the positive aspects. One trick is to follow the 'Hollywood' approach, by presenting the challenge and highlighting the tough road followed, the trials and tribulations of the journey, but

finish by demonstrating how they were overcome. Remember –
everyone wants a happy ending.

While there are clear similarities between poster and oral presentations, there are some nuances that need to be observed when
constructing one or the other. The following sections go into
greater detail on the relative design factors that you should consider when you finally start to put things together on the page.

7.2   PRESENTING POSTERS

Posters summarising the work you have done in your project have
become an increasingly common addition in both undergraduate
and postgraduate courses and provide an opportunity for you to
highlight your creativity. Some institutions also schedule poster
shows that provide an informal setting in which you can talk about
your work to external examiners. In general, there will be a modest
amount of marks associated with your poster creation that will be
split between your actual printed work and your 'defence' of it
upon questioning (usually by two staff members). You do not have
to be an amazing designer to create a good poster, but it is
incredibly easy to produce a bad one. The majority of posters are
prepared using presentation software such as Microsoft Power-
Point or Apple Keynote. Either way you have a single piece of
paper on which to summarise a 50-page dissertation. You already
know what components need to be present, but there are a few
issues that need to be addressed before you start filling the page.

7.2.1   Poster Size

It is likely that you will have been given instructions as to the poster
specification – principally size. The majority of conference posters
are A0, but internal posters tend to be less brash and are usually A1
and occasionally A3. You need to clarify at the outset what the size
limits are so that you are not needlessly penalised. You also need to
determine the orientation of the poster – landscape or portrait.

7.2.2   Poster Templates

A well-organised department probably provides a poster template
that will be a bare PowerPoint slide that has already been

formatted to the correct size and carries the university logo. Again, check on the availability of these. In the absence of a template, consult with your supervisor if there is a particular style that they recommend. A research group can often have its own in-house style and therefore it is wise to consider continuing in that direction.

7.2.3 Poster Background

Remember to keep it simple. There is often a temptation to find a picture that has some relevance to your work and splash it on to the slide as the background. Providing there are no regulations to prevent such use then it is possible to take this approach – but you must be very careful how you use it. While in some cases they can be beneficial, they can complicate matters by making the text difficult to read. Consider the example shown in Figure 7.1A where the introduction to the poster seeks to highlight the need to monitor sulphites in wine.

While the text can still be read in Figure 7.1A, it is far from ideal and becomes a chore. The last thing you want to do is to annoy the examiner. There are several solutions – you could fade the

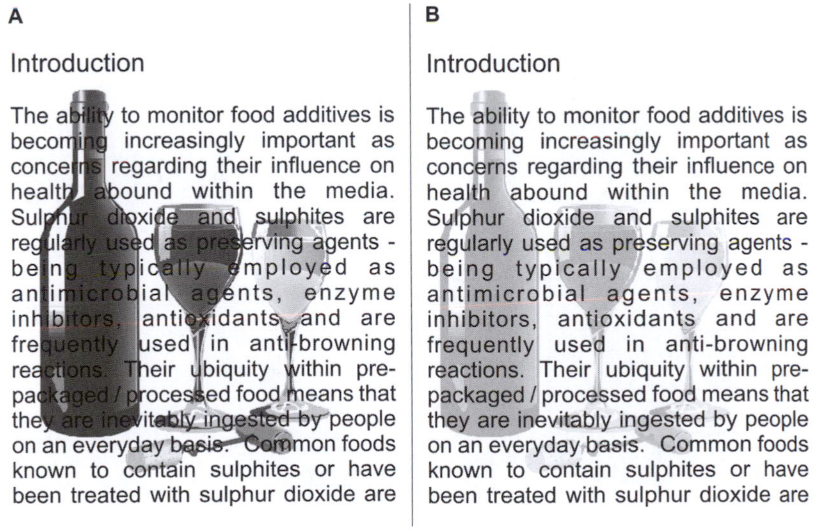

A

Introduction

The ability to monitor food additives is becoming increasingly important as concerns regarding their influence on health abound within the media. Sulphur dioxide and sulphites are regularly used as preserving agents - being typically employed as antimicrobial agents, enzyme inhibitors, antioxidants and are frequently used in anti-browning reactions. Their ubiquity within pre-packaged / processed food means that they are inevitably ingested by people on an everyday basis. Common foods known to contain sulphites or have been treated with sulphur dioxide are

B

Introduction

The ability to monitor food additives is becoming increasingly important as concerns regarding their influence on health abound within the media. Sulphur dioxide and sulphites are regularly used as preserving agents - being typically employed as antimicrobial agents, enzyme inhibitors, antioxidants and are frequently used in anti-browning reactions. Their ubiquity within pre-packaged / processed food means that they are inevitably ingested by people on an everyday basis. Common foods known to contain sulphites or have been treated with sulphur dioxide are

Figure 7.1 Influence of background pictures on text legibility. Before (A) and after (B) picture has been faded.

background such that the picture becomes a faint impression that has only minimal impact on reading the text, but still conveys the context. The alternative is to place the text in boxes or 'panels' that float above the picture. The downside of the latter is that it can obscure the picture and therefore you must ask yourself was it worth putting the picture there in the first place. The situation becomes more complex when importing charts or schematics. They are likely to have a single-colour background that will automatically obscure the picture. Not only that, they will look out of place when next to text that has been directly superimposed on the picture, as highlighted in Figure 7.2.

The use of pictures or coloured backgrounds is really a case of personal choice and it would be wrong to state here that you should follow one particular set style. The use of coloured backgrounds tends to be subject to fashion in that, in the early days, presenters were all too overcome by the ability and opportunity to use colour.

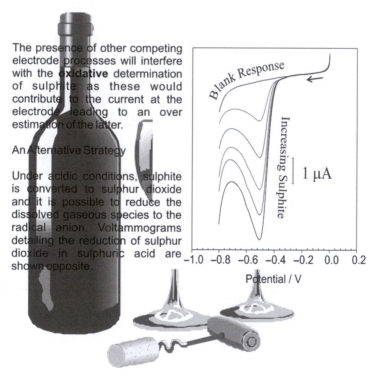

The presence of other competing electrode processes will interfere with the **oxidative** determination of sulphite as these would contribute to the current at the electrode leading to an over estimation of the latter.

An Alternative Strategy

Under acidic conditions, sulphite is converted to sulphur dioxide and it is possible to reduce the dissolved gaseous species to the radical anion. Voltammograms detailing the reduction of sulphur dioxide in sulphuric acid are shown opposite.

Blank Response

Increasing Sulphite

1 μA

−1.0 −0.8 −0.6 −0.4 −0.2 0.0 0.2

Potential / V

Figure 7.2   Effect of importing a JPG chart picture on top of a picture background.

Yes – this is going a fair way back into the mists of time but at one stage colour printing was a novelty! At that point people tended to favour dark background or shaded dark backgrounds – especially blue. There is, however, a question mark over the readability of looking at white text on a coloured background especially as we have become accustomed to reading black text on white paper. The authors personally prefer dark text on white as the contrast is clearer, but we leave it to you to decide which is best for your presentation. It may be that the overall design of your poster requires a dark background and it would be silly to attempt to restrict your creativity.

7.2.4 Font Style

There are hundreds of fonts to choose from but you should restrict yourself to either Arial, Verdana or Calibri. Why? There are two very good reasons for choosing these:

- they are standard on every computer and thus there should be no changes in formatting when it comes to printing off the final version; and
- they are much easier to read than the serif fonts such as Times New Roman or Cursive fonts (that mimic handwriting).

The difference in font structure is highlighted in Figure 7.3. It can be seen that the line thickness in the Arial font is essentially uniform, whereas with the Times New Roman and Brush Script it varies considerably. Consider reading from a distance – the thin strokes are liable to disappear as the text gets smaller. Examine the same text that has been enclosed within the box in Figure 7.3. Which one is easier to read?

Figure 7.3   Influence of font style on legibility.

7.2.5   Font Size

A poster needs to be legible from a distance so you should be thinking in terms of at least Size 16–24 (depending on size of poster) for normal text and 24–30 point for headings. The title will obviously be larger. The font size can be decreased when presenting figures but should not go below 12. If the examiner has difficulty reading the text or the figure axes or annotations then you will lose marks. Play safe and make sure you use a large font size.

7.2.6   General Design Considerations

There are a number of tips that you should follow when designing a poster and these are summarised below

- Make sure your name is on it! You would be surprised how many posters the authors have viewed that fail to mention the name of the presenter! You should also include your student ID number as well.
- Do not try to write a book. Use as little text as possible. You should be able to get an idea of what is going on from a brief glance at the background and the aims. These should be punchy and immediately convey the significance. The last thing you want is to have a poster full of dense text that the viewer has to spend ages reading.
- Have clear margins around the poster. Not only is this better from a design perspective but it is also vital in terms of printing as some printers may not have edge-to-edge capabilities and thus some of your text may not appear! This would be a critical mistake.
- Decide upon a layout – portrait or landscape – then mark out where each section is roughly going to go by placing empty panels onto the page. The size of the panels should reflect the respective sizes of the sections, as indicated in Figure 7.4. Thus, the aims will be relatively small, whereas the results will be the largest. These panels will only be approximate and liable to change once you start adding text but it conquers the fear presented by a blank page. Alter the positions of the panels until you are happy with the layout.
- Start adding text into the panels. Cut and paste from the dissertation and then edit down to get the core points across. Try to work to the panel size you had originally selected.

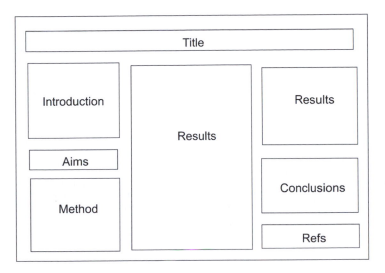

Figure 7.4   Preliminary poster layout.

- Format the text within the panel. If the panels are wide then it may be possible to justify the text so that the edges are even. You have to make a critical judgement as to the readability *vs.* style. While the justified right edge looks neater, it can be annoying where words or the letters within them have been artificially spaced to comply with the formatting. Sometimes it can be better to leave the right-hand side ragged. Try to avoid having panels that are too narrow and can only accommodate one or two words per line. These issues are highlighted in Figure 7.5.
- Paste in the tables and graphs. Check for legibility. Is the text clear and of a size that can be easily read from a metre or two away?
- Avoid adding clipart that does nothing to complement the aims of the project.
- Finally, have you cited the references in the text?

Now stand back and read the poster. Check for typos and cast a critical eye over the design. If you don't like parts of it – chances are that the audience might not either – now is the time to change it. Examples of some poster designs are highlighted in Figure 7.6. Take a moment to consider the design of each. None could be

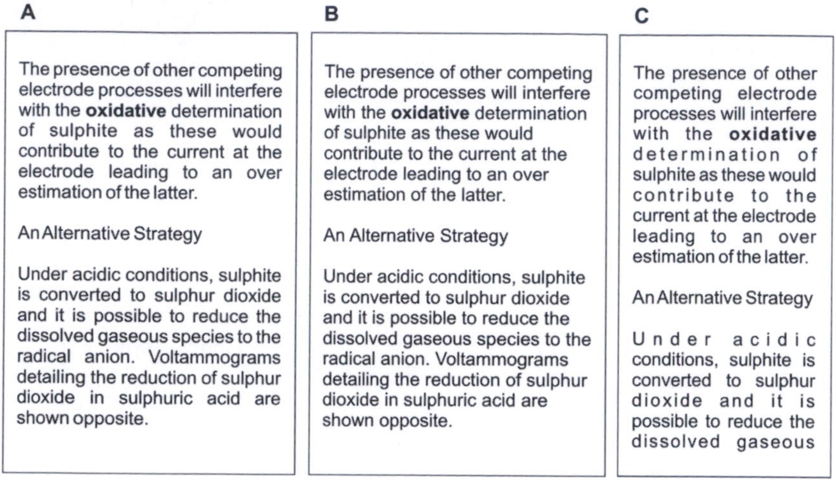

A

The presence of other competing electrode processes will interfere with the **oxidative** determination of sulphite as these would contribute to the current at the electrode leading to an over estimation of the latter.

An Alternative Strategy

Under acidic conditions, sulphite is converted to sulphur dioxide and it is possible to reduce the dissolved gaseous species to the radical anion. Voltammograms detailing the reduction of sulphur dioxide in sulphuric acid are shown opposite.

B

The presence of other competing electrode processes will interfere with the **oxidative** determination of sulphite as these would contribute to the current at the electrode leading to an over estimation of the latter.

An Alternative Strategy

Under acidic conditions, sulphite is converted to sulphur dioxide and it is possible to reduce the dissolved gaseous species to the radical anion. Voltammograms detailing the reduction of sulphur dioxide in sulphuric acid are shown opposite.

C

The presence of other competing electrode processes will interfere with the **oxidative** determination of sulphite as these would contribute to the current at the electrode leading to an over estimation of the latter.

An Alternative Strategy

Under acidic conditions, sulphite is converted to sulphur dioxide and it is possible to reduce the dissolved gaseous

Figure 7.5   Influence of text formatting on legibility.

considered a bad poster – each has it own advantages and limitations. Each tells a story and is easy to follow – that is a crucial component that you must embed in the design you choose.

The posters highlighted in Figure 7.6 differ markedly in style with the top two using a more traditional linear path, whereas the bottom two are much more creative yet are still information rich. If possible try to avoid random positioning of text and chart. The examiner will still expect to be led through the poster rather than having to scan and then attempt to assemble the order in his mind. The bottom two appear to be a random collection, but there is a defined sequence – using either numbers to guide or a top to bottom pathway.

Judging posters can be very subjective. Different people have different ideas of what should go into a poster and the level of detail required. There have been occasions where the authors have attended conferences and seen posters where entire articles have been literally reproduced – there is little chance of anyone wanting to stand and read a journal article spread over an A0 sheet. The key tips are simple and obvious – keep the words to a minimum. Stay away from the gratuitous use of clipart as it invariably looks tacky. Minimise the introduction and emphasise what you actually did and the significance of what you found! Make the design look neat

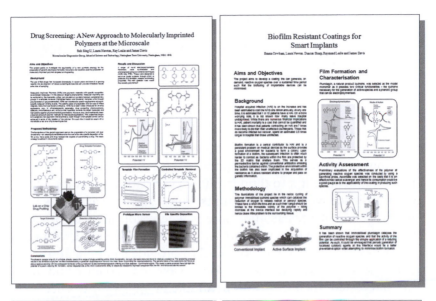

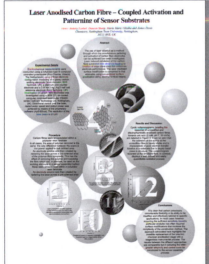

Figure 7.6 Examples of undergraduate project posters.

and tidy and make sure the text is readable and free from typographical or grammatical errors. These are the core issues that will assure you marks, no matter what the examiner thinks about your overall style or how nifty your background is!

7.3   ORAL PRESENTATIONS

Much of the design considerations used in the construction of a poster can also be applied to an oral presentation. There are, however, a number of additional factors to be taken into consideration. The main difference being that instead of presenting the story all at once, you are now going to deliver it bit by bit. When presenting a poster, it is often possible to merge in with the crowd and derive comfort from the anonymity that results. Oral presentations are the complete opposite and the focus is on you. Therefore, in addition to thinking how your slides will come across to the audience, you must also think how you come across as well. Self-analysis is never easy but here is where you need to practice. There are some obvious pointers – you need to be heard, you need to look at the audience and you need to provide a commentary that carries the audience with you. If you engage the audience and make them interested, then they will, in all likelihood, want to cheer you on in the subsequent parts so that you succeed. If they are not interested in the project, then they will not care what comes next and will have, at best, an ambivalent attitude to the outcome. The outcome is that you will score low in terms of marks. These are the three critical factors where students tend to fall down. All too often you will find the presenter talking in a faint mumble to the slides, ignoring the audience with a prepared speech that simply contains what is already on the slides.

Such failings can be remedied, but require a bit of effort on your part and just a touch of willpower in facing down those panic-inducing inner demons when confronted with an audience. First, you need to plan the story. Decide what you want to put in the slides. Follow the story guidelines outlined at the beginning of this chapter. Essentially, you will need a title slide to introduce you, an introduction to set the context, aims to emphasise what your part is, method and then a few results slides and a conclusion/summary slide. It is also a good idea to have an acknowledgement slide at the end to thank all those who helped you in the course of the work. Don't worry too much about the slide design at this stage, just concentrate on the story. It is just like a film – no matter how good the special effects are – it is the story underpinning them that matters. Some communication professionals would argue that you should have a route map slide that effectively sets out the various bits of the talk. . .introduction, aims, method, etc. This is fine where you are engaging in a long presentation, but you only have 10 or 15 minutes and it will become fairly obvious what it is you are presenting – therefore on this occasion it is safe to ignore the presentation route plan.

So script out (long hand or type) what each slide should contain. The hardest part is always the beginning. How do you start? A simple and effective way of introducing your talk is to say something like: "I would like to spend the next 10 or so minutes giving you an overview of the work I did during my project. The project was based on the. ..." This is a simple opening that conveys what you are doing there and what the subject is. There is no need to introduce yourself to the audience – they will know who you are and you will have a title slide with your name on it anyway. The rest of the script is largely up to you. Simply introduce yourself through the title slide, start into the 'overview' sentence and then lead on to the background.

Once you have a rough script, time how long it takes to go through it. Again, don't worry about the slides at this point. A critical mistake is that students put too many words into their commentary, resulting in the talk going over time and generally annoying everyone. Work on the script – honing it till it fits well within the allocated time. Once you have the script you can start to prepare the slides. The words from your script should not end up on the slides. There is no point reading the slide content to the audience. Instead, the slides and commentary should complement each other. In essence, the slides should have minimal text and only cover the main points of what it is you are saying. You fill in the explanation where required by talking directly to the audience. Should you read from the script in front of the audience? No! When you do that, you will isolate yourself from the audience in much the same way as someone who simply turns their back on the audience and reads from the projected slides. Rather, become familiar with what it is you need to say. Rehearsal is vital when conducting your first presentation. When you are familiar with the content of the slides – the trick is to put up the slide – give it a brief glance to ensure that it is in fact the correct one and then use the content on the slide as an aide-mémoire as to what you need to say. The talk itself should be conversational – much as you would talk to your grandmother – though not to your mates in the pub. It is at a level where you convey respect to your audience but maintain an informal style. If you read directly from a script it will, at best, feel and sound stilted. Some people use cards. The slides themselves will function in much the same way as cards, but have the benefit of always being in the correct order. When you are familiar with the work – as you should be – then all you have to do is memorise the opening line to get you going and then explain what the slides represent. The audience will read the slides for themselves whilst listening to you expanding on the content.

What do you do at the end? You can either leave the acknowledgements slide up or have a final slide that is the same as your initial title slide. The last thing you do is present a slide that has 'Questions' in a huge font that spins around and flashes. After presenting the acknowledgement slide – finish the talk by stating something along the lines of: "That essentially summarises the work I have done during my project and I would like to thank you for your attention and would be happy to answer any questions". This is a much more pleasant way of terminating the talk and asking for questions than a garish 'Questions' slide. Remember that you are trying to give a professional talk.

Another point to bear in mind is that you should refrain from moaning during the talk. This is where you may be tempted to state you would have liked to do such and such but couldn't because the instrument was down or you couldn't get access to something, etc. Always try to spin a positive story. If you set out to make mention of something that has limited the pursuit of a project then the audience are justified in asking why didn't you try another approach? Remember, Chapter 2, you are supposed to assess your progress, review the status and then make the changes as required and not simply sit back and say – well it didn't work so I couldn't do any more on it.

Similarly, you need to be judicious in that you want your ending to have a positive spin. Even if things have not worked out quite as you hoped the last thing you want to end on is a downer where you state "tried this – failed, tried that failed...." Rather, use the approach "tried this – the outcome wasn't what we expected and we got this because of...." Here you are spinning something that would otherwise be a negative into a positive through simply viewing the results from a different perspective.

7.3.1 Slide Design

Once you have a script, then you need to prepare slides that contain a summary of what you intend to say. The rules about backgrounds, colour, fonts, etc. all apply here. There are also two other issues that need to be addressed up front: transitions and animations. Please use these sparingly or simply stick to one particular (and unobtrusive) style. There are few things worse than seeing a presentation that tries to use every animation style that PowerPoint will allow. Use the simple 'Appear' or 'Fade' options if you must.

7.3.2 Templates

Size was an important determinant in the construction of a poster but it also has a role in an oral presentation – not in terms of the slide size, but in relation to the number of them. You need to clarify if there is a limit to the number of slides you can present and if there is a departmental or group style template. If not, then it is up to you to choose. The authors would tend to veer towards a simple blank white page – a minimalist style providing a clear contrast in text. However, there are a large number of available templates – some of which may fit in nicely with the context of your project. Try to choose a template that has the design element confined to the border of the slide leaving the majority blank. We would recommend that you shy away from backgrounds where the design encompasses the entire slide – for much the same reasons as those discussed in the Posters section. Two examples of the layout that could be adopted for the presentation are highlighted in Figure 7.7.

The slides in Figure 7.7 are relatively simple in that there is no design clutter from the slide template itself (plain white slide with a fading green band at top and bottom to delineate header and footer). There is minimal text and the slides have schematics that are relevant to the project and that engage the audience. The introduction slide would be presented and the speaker would briefly explain the significance of each of the strategies used in wound management – the slide itself serves to reinforce the main points – which are the different approaches and how they fail to provide long-term antimicrobial action.

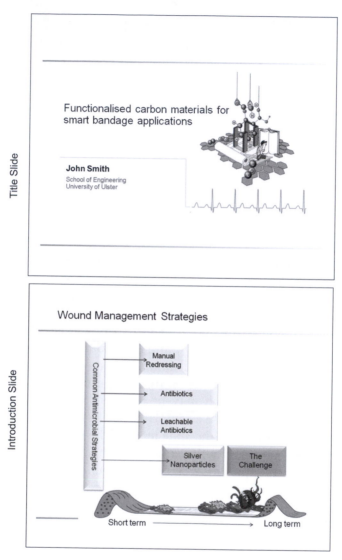

Figure 7.7   Simple slide designs.

7.3.3   Headings

Once you have the number of slides sorted and you've selected your design, you need to put a heading onto each. This must be in a large font and should be a description that immediately shouts to the audience what the slide is about. This is exemplified in Figure 7.7.

The headings should be consistent in terms of style and position through all the slides. The title slide will obviously be the exception.

7.3.4 Content

The main danger when compiling the slide is to add too much text. You inevitably end up shrinking the text to make it fit. There are two issues with this – the first is that nobody wants to read screeds of text and second – if the text is too small – they certainly will not read it. Therefore, when faced with too much text – cut rather than shrink. If a slide is text heavy – then rather than simply confronting the audience with an immediate page of scribble – have the text appear gradually under distinct headings. A slide full of text will quickly give the audience mental fatigue – there is too much to quickly digest therefore they simply wont bother. The solution is to 'drip feed', but remember, you are not to read the text as it appears. When resorting to the drip feed approach – conform to the simple animation of 'Appear' or a rapid 'Fade'. If the animation is slow the audience will be agitated waiting for it to catch up.

An example of the 'drip feed' approach is highlighted in Figure 7.8. The slide is a bit heavy with text but it is important that all the points are covered in one slide. The best answer, if cutting text is not possible, is to have the heading appear – the presenter will explain it and then the next one appear and so on till the message has been delivered. Providing the audience with the complete slide risks overwhelming them and their attention will diminish. The down-pointing arrow is also a indicator that the presenter is leading up to something. This helps to carry the audience as there is a sense of expectation, of piquing curiosity as the audience wants to see what comes next and find out where you are going. The other issue is that when information is littered on the screen – if the audience member breaks concentration to look at the presenter, then they automatically lose their place in the screen text along with the thread of the discussion.

7.4 QUESTION SESSIONS

Most people – staff included – hate questions but the best approach is simply to be honest. That doesn't mean you automatically give in when a gap in your knowledge has been exposed.

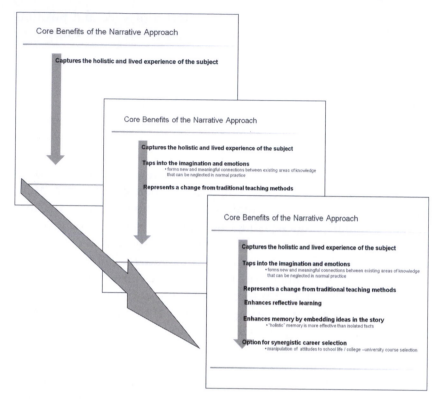

Figure 7.8   Slow release of textual information using animation.

If you are confronted by a question for which you do not immediately have an answer, don't panic and don't say "Sorry, I don't
know". Pause for a minute and try to overcome the terror of "How
do I answer that!" Take a deep breath and think about what was
said. If you think you have an answer, then float it around in your
brain and tease bits out from the darkest corners of your mind. The
staff are not the Inquisition trying to extract information by means
of torture – they are merely trying to develop your ability to think
on your feet. Think it through before uttering a sound – do not be
too hasty to provide an answer – if necessary buy some time by
using the immortal line of "That's a good question – I think…"
then pause – and while you are looking thoughtful try to eek out
the answer. Think about what the question asked. If the question is
long winded then break it down into bits – identify the main terms

used in the question – and use word association to see if you can drag up information from your befuddled brain to help you.

When you have something that you are confident isn't simply mental fluff then open your mouth and probe the questioner with "I think..." and then tentatively add the thought that is now pressing in your mind. Study the face of the questioner and look for the nod or smile as corroboration that you are on the right track. You may well find the person will help you get the answer, but only if you are on the right track and clearly trying. If you are completely lost – then surrender and resort to the "don't know" clause. The best preparation for dealing with questions, however, is to be familiar with the material you are presenting. Are you confident about why you carried out each step? This is where your knowledge of the literature also plays its part as you should be an expert in the work you have done. If there is some element of the project that you do not understand, then you must view it as a weakness that needs to be remedied rather than buried. If you have not studied the literature, there may well be nothing to drag up from your mind and you will be flummoxed.

7.5 SUMMARY/KEY POINTS

☑ Check your course guidelines to determine what size the poster is or if there is a limit on the number of slides.
☑ Check with your supervisor to see if there is a departmental or group template or style.
☑ Try to avoid picture backgrounds on your poster or slide.
☑ Use point 24–26 for the title and 16–18 for the bulk text.
☑ Try to avoid using too much text and where possible add in charts, tables or schematics.
☑ Ensure the text and heading formats are consistent throughout.
☑ Spend time to create a nice design/layout in which the story unfolds.
☑ Remember to clearly specify the aims of the project.
☑ Try to capture the real-world significance of the project.
☑ Ensure that the bulk of the presentation is information rich in terms of results and analysis and not just background.
☑ Remember to include references if you cite prior work.

☑ Most importantly, put your name on the poster or title slide.

☑ When giving an oral presentation – talk to the audience not the screen.

☑ Do not read from the screen.

☑ Ensure that you are heard and don't mumble.

☑ Don't put up a 'Questions' slide.

☑ Don't panic when asked a question – breathe, compose yourself and think before speaking.

CHAPTER 8

Concluding Remarks

8.0 ALMOST THERE

Guidelines? What guidelines?

Research Project Success: The Essential Guide for Science and Engineering Students
Cliodhna McCormac, James Davis, Pagona Papakonstantinou and Neil I Ward
© Cliodhna McCormac, James Davis, Pagona Papakonstantinou and Neil I Ward 2012
Published by the Royal Society of Chemistry, www.rsc.org

The previous chapters are meant to guide you on the journey from selecting a project to the final handing in of the dissertation. There is no suggestion that the advice contained in this book is the definitive guide to producing the perfect dissertation. It would be churlish to suggest that our approach is the best. What we hope to have achieved is the creation of a roadmap or series of guideposts, allowing you to take the time to consider the elements that need to go into your thesis, how to go about finding them along with a consideration of how best to present them. Many students simply don't know what must go in to a thesis, which invariably increases their stress level as deadlines approach and final exams loom on the horizon. We hope the previous chapters will alleviate at least some of that stress through highlighting the critical components that are needed. More importantly, we have tried to indicate the tick boxes that an examiner will use to assess your work. How you present your dissertation is, in the end, very much up to you and we encourage you to be creative and expend some effort on the design and construction of a thesis in which you will be proud. Students start off in a blaze of enthusiasm, but it withers as other pressures mount. Follow the tips outlined previously and try to complete it as you go and you will make it far easier for and on yourself.

8.1 ADDING A FINAL POLISH

If you have followed the advice contained in the previous chapters then you should end up with a professional document that will serve you well in the world beyond the degree course. If you have time, consider adding in a few additional components that will add the final sheen to what should already be a highly polished report. These are supplementary to the main show, but can just add that little bit extra which an examiner will pick up on.

8.1.1 Chapter Dividers

It is a good idea to have some form of divider that separates the main chapters from one another. This could be done by simply inserting an additional 'header' page that has the chapter number and the title centred on the page. This page can easily be incorporated into the manuscript document prior to printing. A better approach, however, would be to print the header separately either

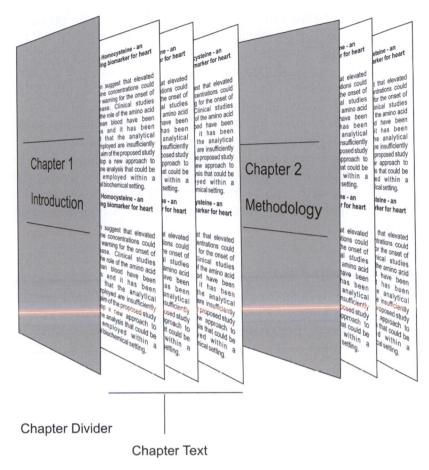

Figure 8.1 Chapter dividers.

on coloured paper or card as indicated in Figure 8.1. If printed separately then the page does not need to be numbered and can be added simply after the other pages have been printed. The advantage of the coloured separator is that it makes it easier for the examiner to locate the different sections and flick between them.

8.1.2 Permission

If you are reproducing diagrams or pictures from printed or electronic sources then you must cite the source of the material

(Chapters 3 and 4) to avoid accusations of plagiarism. In principle, you should ask for permission to reproduce the work from the copyright holder. In many cases this is ignored in undergraduate courses, but it can be worthwhile demonstrating that you are keen to follow correct procedure. The simple approach is to contact the corresponding author on a given publication and send an email explaining your request – make sure that you clarify that it is solely for educational purposes. This may seem like a waste of time, but all it requires is a courteous email that can take 10 minutes to write and send. There is nothing else to do other than wait for a reply. On receipt of a positive reply, you can add 'reproduced with permission' in the figure legend. Place a copy of the correspondence in the appendix.

8.1.3   Newspaper Reports

Scan the general media for stories that are relevant to your project. These can add impact to your poster or oral presentation as they serve to emphasise the real-world significance of the work you have done or are about to do. Add copies of cuttings or a montage if there are many into the appendix.

8.1.4   Bookmark

If your text contains a lot of abbreviations, then consider having a separate pull out section that lists them – in the same style that you used at the front of the dissertation. Make it narrow – about the width of a ruler and attach it to the dissertation by means of a paper clip. It can serve as a useful bookmark for the examiner as they go through the text – the addition of the abbreviation list means they don't have to continually refer back to the list or scan the text for the full description. This is really going the extra mile, but it emphasises the effort you have put into the dissertation.

8.2   FINAL STEPS

The last step is probably obvious. Get someone else to read the dissertation before you submit it. It is difficult to spot typos or errors and your word processing package may not pick them up. All too often, the autocorrect function will covertly attempt to correct a word as you type. It does not always guess correctly and can substitute something that is going to be worse than the original mistake. The misplaced word will fail to show up in a spell check, but can go undetected when you read through as your brain automatically skips along on a presumption of what should be there. You ideally need someone to cast a fresh eye over the content to pick up the silly words that have found their way into the text by mistake, as well as giving you feedback on the overall presentation. In particular, ask them for feedback on the first page of the introduction – remember this is where you are selling the project and if they are not interested in what you have written, then you need to re-write it.

The same is true when preparing for a presentation. We would not necessarily advocate using your family as your focus group as the feedback will invariably be "great, wow, really impressed", which may not be what you need. Ask a friend to listen to your presentation and ideally – go into university early one morning and run through the presentation in the room where it is likely to be held, so that you get a feel for the space.

Finally, despite all the pressure that will fall on your shoulders throughout the final year, try to enjoy the challenges that arise during the project. It is an opportunity to do a little bit of research that no-one else has done and will add to the scientific knowledge base. It will be your contribution and it should be something to be proud of in later years and, who knows... it could lead to greater things.

Subject Index

Page numbers in *italics* refer to entries in figures or tables.